Building Code Basics: Energy

Based on the 2012 International Energy
Conservation Code®

Building Code Basics: Energy

Based on the 2012 International Energy
Conservation Code®

International Code Council
Stephen Kanipe

CENGAGE
Learning·

Australia • Brazil • Japan • Korea • Mexico • Singapore • Spain • United Kingdom • United States

CENGAGE Learning·

Building Code Basics: Energy, Based on the 2012 International Energy Conservation Code®
Stephen Kanipe

Vice President, Technology and Trades Professional Business Unit: Gregory L. Clayton

Director of Building Trades & Transportation Training: Taryn Zlatin McKenzie

Executive Editor: Robert Person

Associate Product Manager: Nobina Preston

Director of Marketing: Beth A. Lutz

Senior Marketing Manager: Marissa Maiella

Production Director: Sherondra Thedford

Content Project Manager: Andrew Baker

Art Director: Benjamin Gleeksman

ICC Staff:

Executive Vice President and Director of Business Development: Mark A. Johnson

Senior Vice President, Product Development: Hamid Naderi

Vice President and Technical Director, Product Development and Education: Doug Thornburg

Director, Project and Special Sales: Suzane Nunes

Senior Marketing Specialist: Dianna Hallmark

For product information and technology assistance, contact us at **Cengage Learning Customer & Sales Support, 1-800-354-9706** For permission to use material from this text or product, submit all requests online at **www.cengage.com/permissions** Further permissions questions can be e-mailed to **permissionrequest@cengage.com**

Library of Congress Control Number: 2012949599

ISBN-13: 978-1-1332-8339-3

ISBN-10: 1-1332-8339-X

ICC World Headquarters
500 New Jersey Avenue, NW
6th Floor
Washington, D.C. 20001-2070
Telephone: 1-888-ICC-SAFE (422-7233)
For all of your building safety needs, visit ICC
Website: http://www.iccsafe.org

Delmar
5 Maxwell Drive
Clifton Park, NY 12065-2919
USA

Cengage Learning is a leading provider of customized learning solutions with office locations around the globe, including Singapore, the United Kingdom, Australia, Mexico, Brazil, and Japan. Locate your local office at: **international.cengage.com/region**

Cengage Learning products are represented in Canada by Nelson Education, Ltd.

Visit us at **www.InformationDestination.com**

For more learning solutions, please visit our corporate website at **www.cengage.com**

Notice to the Reader
Publisher does not warrant or guarantee any of the products described herein or perform any independent analysis in connection with any of the product information contained herein. Publisher does not assume, and expressly disclaims, any obligation to obtain and include information other than that provided to it by the manufacturer. The reader is expressly warned to consider and adopt all safety precautions that might be indicated by the activities described herein and to avoid all potential hazards. By following the instructions contained herein, the reader willingly assumes all risks in connection with such instructions. The publisher makes no representations or warranties of any kind, including but not limited to, the warranties of fitness for particular purpose or merchantability, nor are any such representations implied with respect to the material set forth herein, and the publisher takes no responsibility with respect to such material. The publisher shall not be liable for any special, consequential, or exemplary damages resulting, in whole or part, from the readers' use of, or reliance upon, this material.

Printed in the United States of America
1 2 3 4 5 6 7 17 16 15 14 13

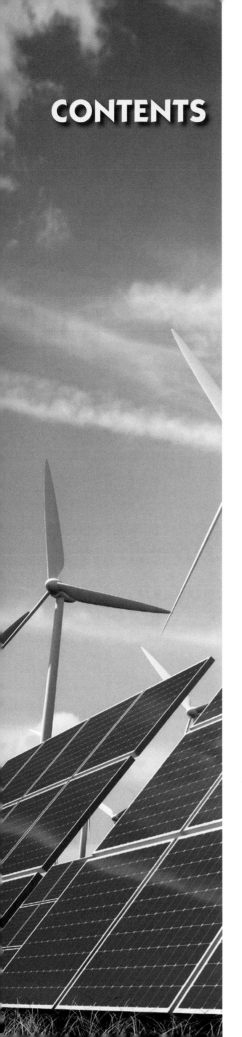

CONTENTS

PART I: 1 INTRODUCTION TO ENERGY AND BUILDING CODES   1

PART II: GENERAL COMMERCIAL ENERGY PROVISIONS   19

v

PART IV: GENERAL RESIDENTIAL ENERGY PROVISIONS   83

PREFACE

Construction practice, building design, and material development changed little over many centuries. People mostly lived in temperate parts of the world, and cold living spaces were managed simply by layering up with heavy clothing and lighting a fire. Hot temperatures were often avoided by changing living habits to be less active during the day and seeking shade and a breeze. Some cultures developed portable shelters to avoid the heat. In all cases and throughout most of history, people built *something* to support their need to live in a safe and comfortable place. Building practice developed over time to suit the local climate and make use of local materials, with the common goal of keeping dust, rain, threatening animals, and bugs out and letting light and fresh air in.

Fire safety has long been a concern as well. As more buildings were constructed in denser city patterns and more people gathered in larger buildings, fire issues had to be addressed. When loss of life and massive property damage become intolerable, the need to regulate construction is satisfied by building codes. The regulatory environment aims to fulfill people's expectation that when they go to work in a building or to an event in a large concert hall, they will breathe clean air and get out safely. The regulations are crafted for the understood need for safety of living and sleeping in the comfort of our homes. The codes adopted by governmental agencies develop over time to refine structural practice so that buildings withstand the forces of nature to safely protect the building occupants and deliver safe, clean water and air to the people inside.

But buildings need to do more than just shelter people and business. As the structural fire resistance and exiting, plumbing, and ventilation systems of buildings became more reliable and the cost of energy increased, building and design professionals began to focus on reducing energy consumption. Regulating energy use in buildings is a relatively new concept in the design and code-enforcement industry. Nationally recognized building safety regulations were published in the early 1900s. The first national energy code was published in the mid-1980s. Code development related to building safety principles has been organized for over a hundred years. Designers, tradespeople, and administrators have had a lot of time to test, try, and teach modern safe building practice. By comparison, the focus on energy efficiency is new to many in the building design and construction trades. The *International Energy Conservation Code* (IECC) developed quickly into a complex document, and the learning curve was steep. The complexity is necessary as innovation in buildings and building systems increase. As new materials, methods, and equipment are introduced, the code provisions change to keep up with advancements. The regulations may overwhelm the homeowner, designer, or builder unfamiliar with the energy code. Sorting through the complex and detailed provisions can be intimidating.

Building Code Basics: Energy captures the provisions that regulate energy use in commercial and residential building construction. It is written to provide a readable and user-friendly overview of the IECC, explaining regulations in clear, noncode language. Understanding energy code provisions is essential to the application of the IECC to any building design; thus, this text is illustrated to further simplify and communicate essential concepts. The text is presented and organized in a user-friendly style with an emphasis on technical accuracy and clear, understandable language. This book is directed to readers familiar with basic construction, architectural, and building system principles but a limited knowledge of energy code requirements and provisions.

Anyone involved in the design, construction, or regulation of building construction can learn from this book. Homeowners, people in nearly any building trade, and those in building design will gain a basic understanding of the principles, provisions, and applications of the technical content of the energy code. The reader will be able to use this book to more fully explore the most common residential and commercial energy efficiency code provisions.

The content of *Building Code Basics: Energy* is organized into commercial and residential building provisions and discusses the administrative regulations that a code official in the building department will use to enforce the energy regulations. The climate zone map and how climate zones affect requirements for different parts of the country are explained, as well as how to use the map. Examples of work that requires a permit and, just as important, work that does not need a permit are identified. Coverage is also provided of the level of detail and information to be included in the building plan documents which are prepared for permit submittal. Requirements for what construction and equipment needs to be inspected and when to call for the inspection are explained. The provisions for basement and wall insulation and discussion about the choices that may best suit a particular condition are introduced in easy-to-follow code references. Requirements for windows, doors, and skylights and their selection and installation per the IECC are covered. The provisions to control air leakage are included in the specific requirements section of the book. Efficient mechanical system specification and installation provisions are included, along with what to do to specify and install a compliant hot water system. Discussion also includes the provisions for electric power and regulations applicable to exterior and interior building lighting.

Even with this extensive coverage, this book is not intended to explain all of the provisions of the commercial and residential energy code or all of the acceptable materials and methods of construction. It focuses on the most common and used provisions applicable to many conditions in residential and commercial construction. This is not to say that the information not covered is any less important or less valuable to the reader. This book should be used with the *2012 International Energy Conservation Code,* which should be referenced for more detail and specific detail.

Reasonable application of the code provisions is supported by a basic understanding of the scope and intent of both the IECC and the other International Codes. This book and the IECC reference other codes and standards, as such a basic understanding of interrelated concepts and provisions provided in this book is necessary.

Building Code Basics: Energy contains full-color photos and illustrations to help the reader visualize and understand the application of the code requirements. Practical examples, simplified tables, and highlights of particularly useful information help clarify the basic requirements of the code and help the reader determine compliance. References to the applicable sections of the 2012 IECC are included to easily locate the applicable code section for more detail and exact code language. A glossary of energy code terms clarifies the meaning of technical terms.

ABOUT THE INTERNATIONAL ENERGY CONSERVATION CODE

The IECC is a comprehensive model code that regulates minimum energy-efficient provisions for new buildings and additions and alterations to existing buildings. There are two separate sets of provisions. The commercial and residential regulations each apply to heating, air-conditioning, ventilation, and lighting systems. Administrative provisions and definitions specific to each commercial and residential set of regulations are also included. The IECC integrates easy-to-understand prescriptive provisions for compliance as well as performance criteria that make possible the use of new materials, new equipment, and new building designs.

The IECC is one of the codes in the family of the International Building Codes published by the International Code Council (ICC). All of these codes are maintained and updated through an open code-development process and are available internationally for adoption by the governing authority to provide consistent and enforceable regulations for the built environment.

ACKNOWLEDGMENTS

Building Code Basics: Energy is the result of many hours of research, code language analysis, and collaborative effort. The author is grateful for the valuable assistance and contributions of Jay A. Woodward, ICC Senior Staff Architect. Jay was the ICC staff secretary to the IECC Code Development Committee during the fast-moving early years of sweeping energy code changes. His unwavering commitment to accuracy and to those actively involved in the tedious code hearing process is much appreciated. The knowledge and experience of the ICC staff were instrumental to its development, and Jay contributed to the accuracy and quality of this product. Hamid Naderi, PE, Vice President of Product Development at the ICC, came up with the concept of this book and provided the initial direction. Thanks to Hamid and Nobina Preston with Cengage Learning for their patience and guidance in preparing the manuscript. Dr. Joseph Lstiburek and Betsy Pettit, F.A.I.A. continually contribute to the ever-growing body of knowledge relating building and energy codes to high-performance building practice. They are among my mentors, and I very much appreciate them and their important work. Finally, thanks to the City of Aspen and the Colorado Chapter of the International Code Council; the city for more than 20 years of financial and in-kind support in cutting-edge local, regional, and international code development and the Colorado Chapter for providing so many opportunities for professional development.

ABOUT THE AUTHOR

Stephen Kanipe, CBO, LEED AP
Chief Building Official
Aspen, Colorado

Mr. Kanipe has 25 years of experience in code administration, plan review, and field inspection. His career started with and continues to be supported by the Aspen Community Development Department. He was appointed to his current position of Chief Building Official in 1995. Mr. Kanipe was selected by the Board of Directors of the International Conference of Building Officials to participate in the International Energy Conservation Code Development Committee and served for five years, including two as chair of that committee. In May of 2009, Stephen was appointed to the ICC's Sustainable Technology Building Committee to help develop the International Green Construction Code (IgCC) and continues to serve on the Energy/Water IgCC Code Development Committee. He is a member of the Colorado Chapter of the International Code Council Past Presidents Committee. Mr. Kanipe was instrumental in developing the Aspen and Pitkin County Renewable Energy Mitigation Program, which was recognized in May 2007 by Harvard University's John F. Kennedy School of Government as a finalist in the Ash Institute Innovations in Government Award. He received an Associate of Science degree in Architectural Technology in 1988 from Columbus State College in Columbus, Ohio. Mr. Kanipe is a Certified Building Official and LEED Accredited Professional.

ABOUT THE INTERNATIONAL CODE COUNCIL

The International Code Council is a member-focused association dedicated to helping the building safety community and construction industry provide safe, sustainable, and affordable construction through the development of codes and standards used in the design, build, and compliance process. Most U.S. communities and many global markets choose the International Codes. ICC Evaluation Service (ICC-ES), a subsidiary of the International Code Council, has been the industry leader in performing technical evaluations for code compliance fostering safe and sustainable design and construction.

Headquarters:
500 New Jersey Avenue, NW, 6th Floor
Washington, DC 20001-2070

District Offices:
Birmingham, AL; Chicago, IL; Los Angeles, CA

1-888-422-7233
www.iccsafe.org

PART

I

Introduction to Energy and Building Codes

Section 1: Energy and Building Code Perspectives

Section 2: Legal Aspects, Code Adoption, and Code Official Authority

© esbobeldijk/www.Shutterstock.com

Energy and Building Code Perspectives

© istockphoto/mustafa deliormanli

INTRODUCTION

Building codes are the various sets of regulations related to the construction, alteration, maintenance, and use of buildings and structures. The codes offer a common and familiar guide for adoption by governmental agencies to approve building plans and inspect construction. The 15 separate codes available include structural design for earthquake, hurricane, and tornado forces; provisions for fire and life safety; provisions for energy conservation; guidelines on systems for heating, cooling, plumbing, and electrical utilities; and guidelines on efficient material, land, and resource use (Figure 1-1). These codes serve primarily to protect the safety and welfare of the building occupants and the public, and to reduce the negative impact of the built environment. The *International Energy Conservation Code* (IECC) provides design guidelines for the effective use and conservation of energy in commercial and residential buildings. The IECC regulates building components in exterior walls such as insulation, windows, and doors and the performance of heating and cooling equipment. The IECC references companion International Codes for elements of construction outside the scope of the IECC, such as mechanical, plumbing, and electrical system installations. This section briefly discusses the history of energy codes and the scope and influence of some of the companion codes and their relationship to the IECC.

HISTORY OF ENERGY CODES

The development of energy codes and standards began in an orderly fashion that continues to be used today. The American Society of Heating, Refrigeration, and Air-Conditioning Engineers (ASHRAE) was founded in 1894 and published Standard 90 in 1975. This standard, titled *Energy Conservation in New Building Design,* was developed in 3 years in accordance with the new National Bureau of Standards *Evaluation Criteria for Energy Conservation in New Buildings.* Standard 90, now ANSI/ASHRAE/IESNA 90.1-2010 (Figure 1-2), is recognized in the 2012 IECC Section C401.2 as a path to compliance for commercial energy efficiency.

Model code organizations followed much the same path developing energy codes in the 1970s. The *Energy Conservation in New Building Construction* code was followed by the *Model Code for Energy Conservation* in 1981, and then the *Model Energy Code* (MEC) in 1983. The MEC (Figure 1-3) was prepared and maintained by the Council of American Building Officials (CABO). This is a significant collaboration in that the MEC brought the then three national code groups—the International Conference of Building Officials (ICBO), the Southern Building Code Council International (SBCCI), the Building Officials Code Administrators International (BOCA), along with the National Conference of States on Building Codes and Standards— together for a common cause: to write an energy code for building officials and the building industry. The U.S. Department of Energy provided the contract to create the 1983 MEC with content based on ASHRAE Standard 90–1980. The document was organized in the common code format and still exists in the structure and organization of the 2012 IECC.

THE CASE FOR BUILDING AND ENERGY CODES

The modern code books are continually developed and improved to help make our built environment safe, healthy, and energy efficient. Code history

FIGURE 1-1 The 2012 I-Codes

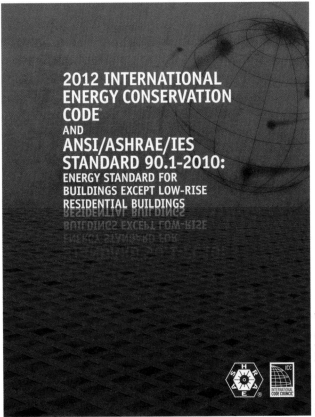

FIGURE 1-2 *International Energy Conservation Code* 2012 Edition, including the ASHRAE 90.1 2010 Edition

© International Code Council

FIGURE 1-3 *The Model Energy Codes* by the Council of American Building Officials

reaches back more than 3,500 years to Hammurabi, whose code created the intent of building regulation based on fairness; construction is to be sound and buildings are to last—consumers should get what they pay for. Fire-resistant construction and improved sanitation principles were implemented as Rome rebuilt after its AD 64 fire. This marked a shift from protecting the building owner and builder to a focus based primarily on risk reduction.

Perhaps as a result of recalling European tragedies, building codes first appeared in the United States in 1625. These early codes were concerned with fire safety and roof coverings. For example, Boston prohibited chimneys made from wood in 1630. Thatch roofs were outlawed at about the same time. George Washington suggested height and area limitations on wood-frame buildings in the District of Columbia. In 1788, the first formal building code in the United States was written in Old Salem, now Winston-Salem, North Carolina.

The cost of nonregulated construction is an issue of national importance. In the summer of 2011, the U.S Congress considered the Safe Building Code Incentive Act (H.R. 2069). Testimony during the hearings stated "Adopting and enforcing newer codes can reduce losses (from natural disasters) by 40 percent or more."

The cost of energy-inefficient buildings is also a matter of national importance. The Department of Energy *Building Energy Codes Resource Guide* published in June 2011 states "Building energy codes are estimated to produce a financial benefit to owners of nearly 2 billion dollars annually by 2015, rising to over 15 billion dollars annually by 2030." In this context, energy costs are a slow-moving disaster. The realized savings are "risk reduction" for building owners and occupants as well as lower utility costs.

CODE DEVELOPMENT

As building systems, materials and mechanical equipment technology continues to develop, the energy code changes to keep up with advancements. This is true for all of the I-Codes; construction technology constantly changes, and the entire suite of codes is revised and published every 3 years. A code change begins as an addition to, deletion of, or improvement to the existing code language. Any interested person or group can submit a code change proposal. The proposed change is first reviewed by an ICC-appointed committee of experts in the construction industry representing contractors, builders, architects, engineers, code administrators, and experts in the specific field of code application. The committees hear testimony from interested parties at a hearing open to the public.

Anyone can speak in support of or in opposition to the proposal. This open debate and broad participation before the committee allows the construction community to fully present and discuss many perspectives on the code change proposal. The committee may approve, modify, or disapprove the code change proposal.

ICC members present at the hearing have the opportunity to object to the vote of the committee. If an objection is sustained by a vote of the membership, the committee action is overturned and the code change proposal moves to the next step. The committee's decisions and results of the hearings are published and made available to all those interested in the code changes. Anyone may submit a written public comment proposing to overturn or modify the committee's decision regarding the original proposal. These comments are assembled and again made available to any and all interested parties. A final-action hearing based on the public comments is then held and, again, the merits of the code change proposal are debated and public comments are welcomed. Only ICC governmental members vote in this last round and can approve, disapprove, or modify the proposal considering the public comments. The results of the final-action hearing and proposals approved by the committee are then incorporated into the next edition of the code.

Code Basics

International Code development cycle

1. Anyone can submit a code change proposal.
2. Proposals are printed and distributed.
3. Open public hearings are held before a committee.
4. Public hearing results are printed and distributed.
5. Anyone can submit public comments on hearing results.
6. Public comments are printed and distributed.
7. An open public final action hearing is held.
8. Final votes are cast by ICC government members.
9. A new edition is published. ●

Code Basics

The code books note the changes with solid vertical lines (|) in the margins, and arrows (→) indicate deletions from the previous edition, as shown below. ●

C403.3 Simple HV AC systems and equipment (Prescriptive). This section applies to buildings served by unitary or packaged HVAC equipment listed in Tables C403.2.3(1) through C403.2.3(8), each serving one *zone* and controlled by a single thermostat in the *zone* served. It also applies to two-pipe heating systems serving one or more *zones*, where no cooling system is installed.

C403.3.1 Ecnomizers. Each cooling system that has a fan shall include either an air or water economizer meeting the requirements of Sections C403.3.1.1 through C403.3.1.1.4.

Exception: Economizers are not required for the systems listed below.

1. Individual fan-cooling units with a supply capacity less than the minimum listed in Table C403.3.1(1).

2. Where more than 25 percent of the air designed to be supplied by the system is to spaces that are designed to be humidified above 35°F (1.7 °C) dew-point temperature to satisfy process needs.

© *International Code Council*

THE BUILDING CODES: SCOPE AND LIMITATIONS

A number of features are common to all of the International Codes. Each code begins by stating its scope of application. The scope establishes the range of buildings, uses, construction, equipment, and systems to which the code applies. A purpose statement follows the scope and includes the intent to provide minimum standards to protect the health, safety, and welfare of the public. Subsequent sections place limitations on the application of the code. For example, each code permits the continued legal use of existing buildings if the buildings do not create hazards to the occupants or property and meet certain minimum acceptable standards for health and sanitation; such buildings do not need to be brought into compliance with the current codes. In the case of an addition to an existing building, for instance, only the addition need comply provided it does not cause an unsafe condition in the existing structure. Each International Code also references other codes and standards for use under specific circumstances. For example, the IECC references the *International Mechanical Code* (IMC) for installation of heating, cooling, and ventilation equipment. It also has requirements for public health and safety such as to remove automobile automobile exhaust from a parking garage. These provisions are beyond the scope of the IECC. The efficiency and performance of the equipment is within the scope of the IECC. Together, the 15 codes provide a comprehensive body of regulations to be adopted and enforced to provide citizens a minimum acceptable level of protection of life and property from fire and other hazards and adequate light, ventilation, and egress. Finally, the appendices of each code are not in effect unless they are specifically adopted by the local authority jurisdiction.

International Building Code (IBC)

The provisions of the *International Building Code* (IBC) apply to the construction, alteration, maintenance, use, and occupancy of all buildings and structures, (Figure 1-4) except those covered by the *International Residential Code* (IRC). In addition to structural components and systems, the IBC provides for a safe means of egress, accessibility for persons with disabilities, fire resistance, fire protection systems, weather resistance, and finishes and interior environments. These regulations are typically related to the use and occupancy of the building. That is, the IBC assesses relative risks or hazards based on the functions within the building and controls design accordingly. Provisions regulating

FIGURE 1-4 *International Building Code* 2012 Edition

FIGURE 1-5 A community college building regulated by the IBC

FIGURE 1-6 A single family home regulated by the IRC

the building's size, means of egress, fire-resistive elements, and fire protection systems vary significantly among different types of buildings, for example sports arenas, hospitals, schools (Figure 1-5), apartments, and office buildings.

INTERNATIONAL RESIDENTIAL CODE

The *International Residential Code* (IRC) regulates the construction of one- and two- family dwellings and town home structures. These particular buildings are not included in the scope of the IBC. The IRC provisions include all of the regulations for building structural elements, fire and life safety, mechanical, fuel gas, plumbing, electrical and energy in one document. This creates a stand-alone code for residential construction that provides strong, stable, and sanitary homes that conserve energy while still offering adequate lighting, comfort conditioning and ventilation. The IRC is one source of regulation for homeowners, home builders and jurisdictions. It provides many prescriptive provisions for construction of homes (Figure 1-6).

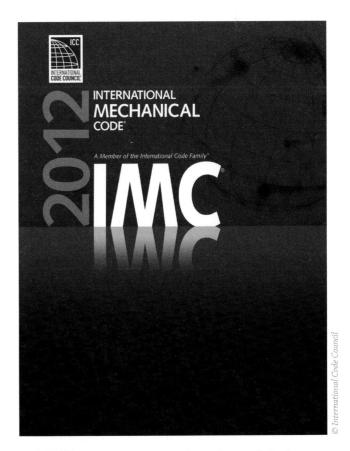

FIGURE 1-7 *International Mechanical Code* 2012 Edition

International Mechanical Code (IMC)

The provisions of the *International Mechanical Code* (IMC) generally apply to the installation, alteration, use, and maintenance of permanent mechanical systems utilized for comfort heating (Figure 1-7), cooling, and ventilation (HVAC), and other mechanical processes within buildings (Figure 1-8).

FIGURE 1-8 Hotel mechanical equipment installation regulated by the IMC

International Fuel Gas Code (IFGC)

The *International Fuel Gas Code* (IFGC) regulates the installation of natural gas and LP-gas piping systems, fuel gas utilization equipment, gaseous hydrogen systems, and related accessories (Figure 1-9). The fuel gas piping system extends from the utility company point of delivery to the equipment shutoff valves. Code coverage includes pipe sizing and other design considerations, approved materials, installation, testing, inspection, operation, and maintenance. The equipment installation requirements include combustion and ventilation air, approved venting, and connection to the piping system.

International Plumbing Code (IPC)

The provisions of the *International Plumbing Code* (IPC) generally apply to the installation, alteration, use, and maintenance of plumbing systems (Figure 1-10). The IPC includes the material and installation requirements for water supply and distribution, plumbing fixtures, drain waste and vent piping, and storm drainage systems.

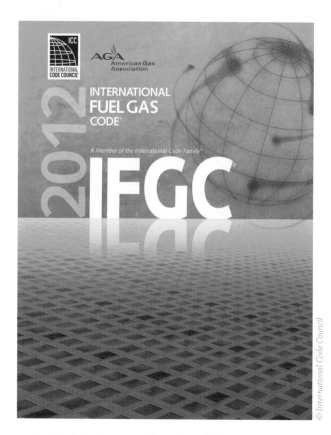

FIGURE 1-9 *International Fuel Gas Code* 2012 Edition

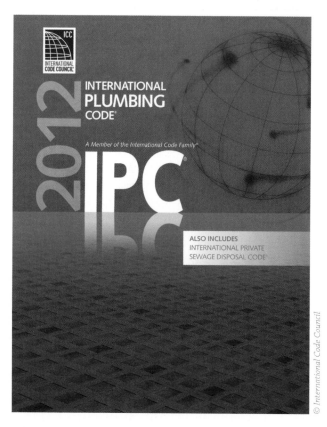

FIGURE 1-10 *International Plumbing Code* 2012 Edition

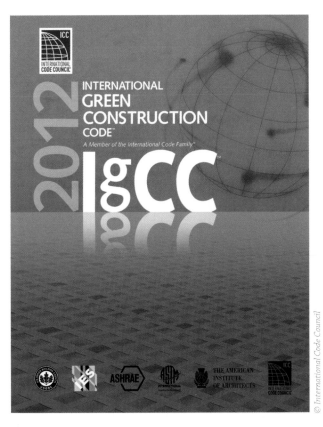

FIGURE 1-11 *International Green Construction* Code 2012 Edition

FIGURE 1-12 High rise residential building with installed wind powered electric generation

International Green Construction Code (IgCC)

The *International Green Construction Code* (IgCC) is the newest member of the I-Code family. It has been developed to safeguard the environment, public health, and the general welfare of people, places, and resources affected by development and buildings. (Figure 1-11) The provisions intend to reduce the negative impacts and increase the positive impacts (Figure 1-12) of the built environment on the natural environment and building occupants.

INTERNATIONAL ENERGY CONSERVATION CODE (IECC)

The *International Energy Conservation Code* (IECC) intent statement is curiously different than those of the other codes and does not speak specifically to life safety and property protection (Figure 1-13). In fact, the statement points out that the IECC is not intended to reduce any safety, health, or environmental requirements in other applicable codes or ordinances. The intent of the IECC is to "regulate the design and construction of buildings for the effective use and conservation of energy over the useful life of each building." Further, the code is intended to provide flexibility to permit the use of innovative approaches and techniques to achieve this objective.

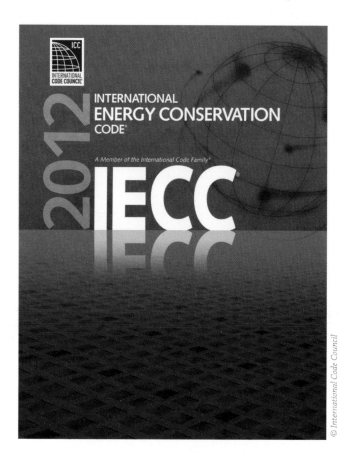

© International Code Council

FIGURE 1-13 *International Energy Conservation Code* 2012 Edition

It is this divergence from life safety and property protection that makes the energy code unique in the family of International Codes. Many involved in the design, construction, and inspection of buildings recognize that now is the time to begin to understand the energy code. The 2012 IECC meets the need for a design and construction document in common code format to address energy-efficient building enclosures and mechanical, lighting, and power systems.

The IECC, as well as *Building Code Basics: Energy—Based on 2012 International Energy Conservation Code,* acknowledges distinct differences in requirements for commercial and residential buildings. Commercial and residential requirements are organized in separate provisions of the 2012 IECC. The requirements for permits and inspections, climate zones, and code administration are almost the same for commercial and residential construction, but insulation and mechanical requirements greatly differ. Even the definitions used to establish the common vocabulary for commercial and residential energy requirements vary. The commercial provisions are all in six sections of IECC Section C4 and are included in Part III of this book. IECC Section R4 contains the residential provisions. These requirements are addressed in Part V of this book.

Legal Aspects, Code Adoption, and Code Official Authority

© International Code Council

Building codes establish minimum regulations intended to benefit citizens while living in, working in, or visiting structures in the built environment. Most model code provisions focus on the health and safety of the occupants. The intent of the International Energy Conservation Code (IECC) is to reduce building energy use and increase the comfort of our homes and offices. Just like the "life safety" codes—building, plumbing, and mechanical—the IECC must be legally adopted to become law. This section describes the adoption process and explores local amendments required for effective administration and enforcement of the energy code provisions.

CODE ADOPTION

The Constitution of the United States of America grants states jurisdiction over regulation of building construction. Each state operates by specific statutes legislating authority regarding the adoption of building codes. Some states regulate building construction by appointing a specific board or agency. The statutes of many states allow the local county, city, or recognized governmental organization to adopt a code to regulate building construction. The IECC is designed to be adopted by reference by ordinance. A sample of typical legislation is included in the introductory pages of the IECC (Figure 2-1). The adoption must include the specific local information in the bracketed locations in the sample figure.

Changes to the International Codes are made every 3 years, and the updated versions, especially in the field of building energy use, recognize new materials, developing technology, and efficient and effective methods of construction. It is important for homeowners, designers, and contractors to be familiar with all the codes and amendments adopted and enforced in the jurisdiction in which their project is being designed and built. This action is a necessary first step in the building permit process.

Amending the IECC

Model codes provide the advantage of consistency and uniformity in construction code adoption in a jurisdiction and across jurisdictional boundaries. Uniformity is good for material and equipment manufactures, designers, builders, and code officials in charge of administering and enforcing building regulations. It is a fact that every model code adoption requires amendments to reflect specific local conditions. Some states and jurisdictions incorporate many amendments in the adopting ordinance and others very few changes to the code text as published.

The energy code may be one of the codes a jurisdiction chooses to amend. This is necessary to acknowledge unique local climate and weather conditions that do not agree with the broader climate zones published in the IECC. Amendments are also driven by politics, local building tradition and practice, and required compatibility with other state or local laws. Any amendment included in the adopting ordinance is part of the law and must be recognized by all users regarding the energy code provisions.

Codes and Standards

You Should Know

Chapter 5 of the 2012 IECC provides the listing of Referenced Standards. Applicable standards of organizations such as ASTM International, ASHRAE and CSA are included. ●

Codes and standards are sometimes mentioned in the same sentence and may be thought to be very similar documents. There is a difference, however, and code users should know the difference. A code is a body of laws that helps a city, state, or country establish rules of compliance. The Code of Hammurabi, for example, is a well-known ancient set of laws, and law 229 speaks directly to building regulation: *If a builder builds a house for someone, and does not construct it properly, and the house which he built falls in and kills its owner, then the builder shall be put to death.* This is a performance code. The house should not fall in, but provisions are not offered to guide the builder in exactly how to prevent this occurrence. Prescriptive code language is a provision that details exact requirements

SAMPLE LEGISLATION FOR ADOPTION OF THE *INTERNATIONAL ENERGY CONSERVATION CODE* ORDINANCE NO. _____

A[N] [ORDINANCE/STATUTE/REGULATION] of the **[JURISDICTION]** adopting the 2012 edition of the *International Energy Conservation Code,* regulating and governing energy efficient building envelopes and installation of energy efficient mechanical, lighting and power systems in the **[JURISDICTION];** providing for the issuance of permits and collection of fees therefor; repealing **[ORDINANCE/STATUTE/REGULATION]** No. _____ of the **[JURISDICTION]** and all other ordinances or parts of laws in conflict therewith.

The **[GOVERNING BODY]** of the **[JURISDICTION]** does ordain as follows:

Section 1. That a certain document, three (3) copies of which are on file in the office of the **[TITLE OF JURISDICTION'S KEEPER OF RECORDS]** of **[NAME OF JURISDICTION],** being marked and designated as the *International Energy Conservation Code, 2012* edition, as published by the International Code Council, be and is hereby adopted as the Energy Conservation Code of the **[JURISDICTION],** in the State of **[STATE NAME]** for regulating and governing energy efficient building envelopes and installation of energy efficient mechanical, lighting and power systems as herein provided; providing for the issuance of permits and collection of fees therefor; and each and all of the regulations, provisions, penalties, conditions and terms of said Energy Conservation Code on file in the office of the **[JURISDICTION]** are hereby referred to, adopted, and made a part hereof, as if fully set out in this legislation, with the additions, insertions, deletions and changes, if any, prescribed in Section 2 of this ordinance.

Section 2. The following sections are hereby revised:

Sections C101.1 and R101.1. Insert: **[NAME OF JURISDICTION]**.

Sections C108.4 and R108.4. Insert: **[DOLLAR AMOUNT]** in two places.

Section 3. That **[ORDINANCE/STATUTE/REGULATION]** No. _____ of **[JURISDICTION]** entitled **[FILL IN HERE THE COMPLETE TITLE OF THE LEGISLATION OR LAWS IN EFFECT AT THE PRESENT TIME SO THAT THEY WILL BE REPEALED BY DEFINITE MENTION]** and all other ordinances or parts of laws in conflict herewith are hereby repealed.

Section 4. That if any section, subsection, sentence, clause or phrase of this legislation is, for any reason, held to be unconstitutional, such decision shall not affect the validity of the remaining portions of this ordinance. The **[GOVERNING BODY]** hereby declares that it would have passed this law, and each section, subsection, clause or phrase thereof, irrespective of the fact that anyone or more sections, subsections, sentences, clauses and phrases be declared unconstitutional.

Section 5. That nothing in this legislation or in the Energy Conservation Code hereby adopted shall be construed to affect any suit or proceeding impending in any court, or any rights acquired, or liability incurred, or any cause or causes of action acquired or existing, under any act or ordinance hereby repealed as cited in Section 3 of this law; nor shall any just or legal right or remedy of any character be lost, impaired or affected by this legislation.

Section 6. That the **[JURISDICTION'S KEEPER OF RECORDS]** is hereby ordered and directed to cause this legislation to be published. (An additional provision may be required to direct the number of times the legislation is to be published and to specify that it is to be in a newspaper in general circulation. Posting may also be required.)

Section 7. That this law and the rules, regulations, provisions, requirements, orders and matters established and adopted hereby shall take effect and be in full force and effect **[TIME PERIOD]** from and after the date of its final passage and adoption.

FIGURE 2-1 Example of an adopting ordinance

required for compliance, such as Chapter R402.2 of the IECC, "Specific Insulation Requirements," which addresses ceilings, walls, slabs, and other building components with information on exactly what the builder shall do to comply with the law. When "shall" is used in a building code provision the regulation is mandatory.

Standards are developed to establish a target of performance. Materials and methods are measured and tested to determine performance and compliance to the applicable standards. This issue is so important

Code Basics

Jurisdictional amendments become part of the code provisions and are required to be made available to permit applicants doing work there. Know the amendments in the jurisdiction having authority to review plans and issue the permit. The following is an example of amended provisions specific to IECC inspections:

(a) Section C104.3 "Final Inspection" is hereby amended and to read as follows:

Section C104.3 Required energy efficiency inspections. The *building official*, upon notification, shall make the inspections set forth in Sections C104.3.1 through C104.4.5.

Section C104.3.1 Building thermal envelope. Specific items referenced in Section C402.4.1 shall be inspected and approved.

Section C104.3.2 Fenestration. All fenestration as defined in Section C202 shall be labeled, inspected and approved before the gypsum board inspection.

Section C104.3.3 Insulation. All above and below grade wall and ceiling cavity insulation shall be labeled, inspected and approved before the gypsum board inspection.

Section C104.3.4 Other inspections. In addition to the inspections specified above other inspection shall include, but not be limited to, inspections for: duct system insulation R-value, HVAC and water-heating equipment efficiency and lighting compliance.

Section C104.3.5 Final inspection. The building shall have a final energy efficiency inspection and not be occupied until *approved*. ●

that referenced standards get their own code sections and Chapter in the building codes. It is generally easy to connect code language and standard applicability. Chapter 5 in both the commercial and residential code provisions lists the referenced standard with the publication date, edition year, title, and code section referencing it (Figure 2-2).

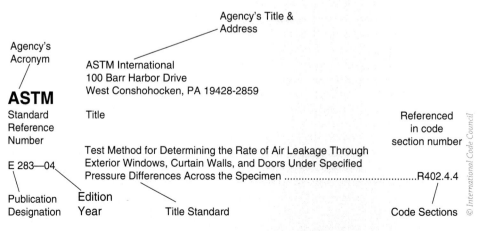

FIGURE 2-2 Referenced standard example

The specific standard information provided to demonstrate code compliance must agree with the standard in Chapter 5. The specific code provision applies when there is a conflict between the code and the standard.

Federal Law

Energy efficiency is near the top of the national agenda. The potential economic benefits realized in building-related manufacturing, product development, and employment for a skilled workforce are incorporated in the American Reinvestment and Recovery Act of 2009 (ARRA). This legislation provided funds to states to educate code officials, builders, designers, engineers, and public officials regarding the positive economic impacts of adopting the IECC.

The ARRA provisions reference the 2009 IECC provisions for residential buildings and ANSI/ASHRAE/IESNA Standard 90.1-2007 for commercial buildings, or equivalent codes, to establish effective and efficient building performance regarding energy use. The ARRA, signed on February 14, 2009, includes provisions adopting the most recent edition of the energy code. This is important because ARRA requires that states demonstrate 90 percent compliance by 2017 as a condition of receiving the funds. As the 2012 IECC has been determined by the Department of Energy as more efficient for both residential and commercial construction, adoption of the most recent edition of the IECC moves states closer to compliance with ARRA. As a result of the ARRA, the federal government recognizes two documents, the 2009 IECC and ASHRAE 90.1-2007, as benchmarks for energy efficiency.

You Should Know

Several federal agencies are actively involved in building safety and energy efficiency arena, such as the Department of Energy (DOE), Department of Housing and Urban Development (HUD), Department of Justice (DOJ), National Institute of Standards and Technology (NIST), and Federal Emergency Management Agency (FEMA). ●

AUTHORITY

The IECC grants the code official authority to enforce the requirements in the energy code. When it is adopted, the code becomes law and, like any other law, must be enforced. The code official establishes a fee schedule and cannot issue the permit until all the fees are paid. Work that requires a permit may not start before the permit is issued. The code official has the authority to stop work that has not been approved and fine the person if the order is ignored and the work continues.

The code official also establishes a board of appeals to hear problems permit holders may have with decisions or orders made concerning the application and interpretation of the energy code. This board hears such an appeal, makes a decision, and delivers it in writing to the appellant and code official. Thus, the code grants broad authority to the code official but creates the board of appeals to assure the authority does not go unchecked.

DUTIES OF THE CODE OFFICIAL

The code official is charged with issuing plans reviewed for code compliance and inspecting the construction for compliance with the approved plans and the energy code. The code official may use several methods to achieve this primary function. Alternate materials, methods of construction, or design or insulating systems may be used when approved by the code official. Other energy code programs may be considered in compliance with the IECC if the code official of the jurisdiction determines that the alternate program exceeds the energy efficiency requirements of this code.

Plan Review

The code official also has the job of accepting plans and documentation for review and performs this duty by requiring that information specific to energy compliance must be included in the building plans. The documents must clearly show the where, what, and how of the work that is proposed for the building project. If the details, sections, insulation, windows, mechanical design, ductwork, and lighting plan do not sufficiently include the information needed for energy code review, the code official may send the applicant back to the drawing board. The plans are submitted and reviewed before construction so that noncompliant details or conditions can be corrected prior to construction to save costs and avoid problems after a structure has been built (see Figure 2-3).

An ICC-ES report provides information required to approve the specified product (Figure 2-4). The report describes the product, details installation instructions, lists specific conditions of use and identification of a compliant product.

When the code official accepts project plans and documentation, a plan review for compliance with the energy code requirements in addition to a review of building life safety, mechanical, plumbing, and electric codes must be performed. When the code official determines that the plans comply with all the applicable codes, the permit documents are stamped "REVIEWED FOR CODE COMPLIANCE" and issued (Figure 2-5). The code official keeps a set and another is kept on the job site in the field. These approved plans are for the builder to follow and the code official to reference during inspections (Figure 2-6).

Inspections

Any work or construction that requires a permit also requires inspection. The extent and complexity of the project determines the construction process and therefore the inspection sequence. This is so because the energy code does not allow work to continue beyond what has been inspected and approved by the code official. Specific inspections for wall, ceiling, and floor insulation, windows, duct insulation, and mechanical and water heating efficiency are required by the *International Building Code* (IBC).

© International Code Council

FIGURE 2-3 Plans reviewer at work

 **ICC EVALUATION SERVICE**

Most Widely Accepted and Trusted

ICC-ES Evaluation Report

ESR-1231

Reissued March 1, 2009
This report is subject to re-examination in one year.

www.icc-es.org | (800) 423-6587 | (562) 699-0543 *A Subsidiary of the International Code Council®*

DIVISION: 09—FINISHES
Section: 09270—Gypsum Board Accessories

REPORT HOLDER:

EZ TAPING SYSTEMS, INC.
POST OFFICE BOX 11263
GREEN BAY, WISCONSIN 54307-1263
(920) 429-9274

EVALUATION SUBJECT:

EZ TAPING SYSTEM "FIRE TAPE" DRYWALL TAPE

1.0 EVALUATION SCOPE

Compliance with the following codes:

- 2006 *International Building Code®* (IBC)
- 2006 *International Residential Code®* (IRC)
- 1997 *Uniform Building Code™* (UBC)

Property evaluated:

Fire resistance

2.0 USES

The assembly described in Section 4.2 of this report can be used where a one-hour nonload-bearing fire-resistance-rated steel- or wood-framed assembly is required by Chapter 7 of the IBC, Section R317 of the IRC or Chapter 7 of the UBC. When used in this assembly, the EZ Taping System Fire Tape is an alternative to the joint treatment required by IBC Section 2508.4, IRC Section R702.3.1 and UBC Section 2511.5.

3.0 DESCRIPTION

The EZ Taping System Fire Tape is a self-adhesive, fiberglass-reinforced, paper drywall tape. The tape is 1.9 inches (48 mm) wide and comes in rolls 250 feet (76.2 m) long.

4.0 INSTALLATION

4.1 General:

The tape is applied directly to vertical or horizontal joints of gypsum wallboard panels. The tape must be wiped down with a 3-inch (72 mm) plastic blade provided by the manufacturer or with an equivalent instrument to apply sufficient pressure to establish full contact between the tape and drywall.

4.2 One-hour Nonload-bearing Fire-resistance-rated Assembly:

The construction consists of minimum 0.0185-inch-thick (0.46 mm), $3^5/_8$-inch-deep (92 mm) steel studs and tracks or 2-by-4 wood studs faced on both sides with $^5/_8$-inch-thick (15.9 mm), Type X gypsum wallboard. Studs are spaced a maximum of 24 inches (610 mm) on center. The wallboard is erected vertically or horizontally and attached to studs and tracks with $1^1/_2$-inch-long (38 mm), Type S drywall screws for steel studs, or Type W drywall screws for wood studs, spaced 8 inches (203 mm) on center on vertical edges and 12 inches (305 mm) on center on top and bottom edges and in the field. The wallboard panels must be tightly butted, with all joints blocked. Joints must be centered on the stud face and staggered one stud on opposite faces of the assembly. The Fire Tape wallboard tape is applied to the wallboard joints in accordance with Section 4.1. Fasteners shall be treated with approved joint tape or joint compound.

5.0 CONDITIONS OF USE

The EZ Taping System described in this report complies with, or is a suitable alternative to what is specified in, those codes listed in Section 1.0 of this report, subject to the following conditions:

5.1 The tape is manufactured, identified and installed in accordance with this report and the manufacturer's instructions.

5.2 The tape is applied to tightly butted joints of nonload-bearing one-hour fire-resistive assemblies as described in Section 4.2.

6.0 EVIDENCE SUBMITTED

6.1 Reports of tests in accordance with ASTM E 119, C 474 and C 475, and the ICC-ES Acceptance Criteria for Adhesively Attached Drywall Tape (AC119), dated July 1996.

6.2 A quality control manual.

7.0 IDENTIFICATION

Each roll of tape is marked with the manufacturer's name (EZ Taping System) and the evaluation report number (ESR-1231). In addition, each package of product contains installation instructions, limitations on use and a plastic knife for use in applying the tape.

FIGURE 2-4 ICC evaluation report

FIGURE 2-5 Approved plans stamp

To illustrate this, if a basement wall is approved with exterior insulation, the area must not be backfilled before the below-grade insulation inspection is called for and approved. Work that does not comply must not be covered up until approved to do so by the code official. Likewise, if a wall cavity insulation inspection fails, the corrections must be made and the inspection approved before the wall finish material is applied to the studs.

Tests and inspections specific to the energy code must be done to verify the work complies with the IECC. If the work does comply, the code official issues a notice of approval. This is usually a note or box on the inspection request to be marked "Approved" or "Rejected". The final inspection must be called for and approved before the building is occupied.

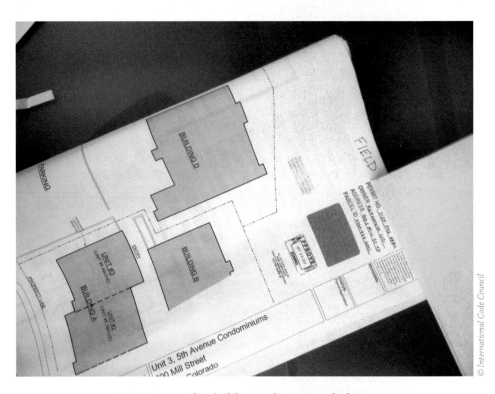

FIGURE 2-6 The field set of approved plans

PART

II

General Commercial Energy Provisions

Section 3: General Commercial Energy Provisions

Section 4: Administration and Enforcement

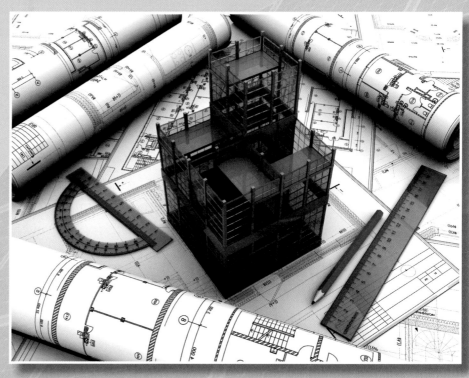

© Mmaxer/www.Shutterstock.com

General Commercial Energy Provisions

© International Code Council

It is necessary to establish a framework to address the administration, application, and enforcement of the commercial energy provisions set forth in the *International Energy Conservation Code* (IECC), *International Residential Code* (IRC), and *International Building Code* (IBC), among other codes. The code official's responsibilities regarding plan review and inspections, the architect's and engineer's responsibilities concerning document preparation, and the contractor's responsibilities in regard to permits and inspections are clearly stated in the IECC. Code sections C101 and C102 of the IECC, discussed in this section, govern the relationships and understandings between the building department authority and the design and construction community.

SCOPE

These code provisions apply to commercial buildings and building sites, systems, and equipment. An italicized term in the IECC means the word or phrase has a specific meaning in code language and that a specific definition is used to clarify the meaning of that term. Chapter 2 of the IECC lists 85 words and specific definitions to establish the common vocabulary for the commercial energy regulations. *Commercial buildings* in this code are defined as "all buildings that are not included in the definition of *residential building*." Although this may not seem to provide enough information, those familiar with codes recognize this as a pointer to the definition of *residential building*. In the IECC, a residential building "includes detached one- and two-family dwellings and multiple single-family dwellings (townhouses) as well as group R-2, R-3, and R-4 buildings three stories or less in height above grade plane. **[Ref. 202]** This definition is specific to the IECC and is different than the definitions in the IRC and IBC.

As another example, the term *story above grade plane* can be defined by working through several definitions in the IBC. A story more than 6 feet above the average grade around the exterior wall of the building or more than 12 feet above grade at any point is considered a *story above grade plane* (Figure 3-1). A *story* is defined as "that portion of a building included between the upper surface of a floor and the upper surface of the floor or roof next above." The *grade plane* is "a reference plane representing the average of finished ground level adjoining the building at *exterior walls*." This measurement is used to determine building height. A basement is a story that is mostly below finished ground level.

Classifications of residential occupancies, which are defined as units where people live, eat, and sleep, are found in the IBC. Apartment and condominium buildings are multifamily structures and represent a typical R-2 occupancy. Common R-4 buildings are group homes, small assisted-living facilities, and halfway houses. The shared attribute of these residential uses is that the occupants are "non-transient." A three-story apartment building is not regulated by the commercial energy provisions (Figure 3-2). A three-story hotel is an R-1 building and must comply with the commercial energy provisions.

Thus, if the building uses do not fit into any of the defined descriptions of "residential" or the building is more than three stories in height, the IECC commercial provisions must be applied. Common commercial uses are offices, banks, clothing stores, restaurants, bars, retail sales, automobile

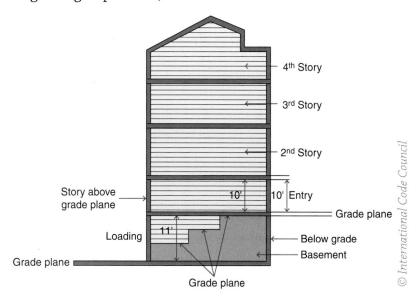

FIGURE 3-1 Story above grade plane

© International Code Council

FIGURE 3-2 A three-story apartment building is not regulated by the commercial energy provisions

repair shops, and gyms. The IBC commercial occupancy groups are Assembly, Business, Educational, Factory, High-hazard, Institutional, Mercantile, Residential, Storage and Utility, and Miscellaneous. A four-story hotel (Figure 3-3), tall office building (Figure 3-4), fire station (Figure 3-5), and corner store (Figure 3-6) are all regulated by the commercial energy code provisions. **[Ref. C101.5]**

INTENT

The intent of the IECC is stated simply in the code: "This code shall regulate the design of and construction of buildings for the effective use and conservation of energy over the useful

FIGURE 3-3 Four-story hotel

FIGURE 3-4 Office building

FIGURE 3-5 Fire station

FIGURE 3-6 Corner store

life of each building." The energy code regulations provide options for design and construction to accomplish the intent to conserve energy use. These regulations apply to new buildings and building systems, as well as construction projects in existing buildings. "Over the useful life of each building" is a somewhat vague statement, as the "useful life" is difficult to determine, and affected by many factors. Whereas some buildings are only meant for a useful life of 50 years or so, most government, university, and hospital buildings are designed to be useful for hundreds of years. Typically, the longer the intended useful life of a building, the more it costs to build.

Although the IECC lists the most commonly used and standard methods, it is not meant to limit the use of alternative approaches, equipment, or techniques that also conserve energy. If a mechanical, plumbing, or electrical system; insulation material; or building envelope technique is not specifically listed as allowed in the code or does not meet the strict letter of the code, it may nonetheless be allowed by the building official, as energy-conserving innovations are encouraged. For example, Figure 3-7 illustrates the use of straw bale as insulation, and Figure 3-8 shows straw bale used as a building material for a school. [Ref. C101.3]

FIGURE 3-7 Straw bale insulation

FIGURE 3-8 Straw bale as a building material

APPLICABILITY

The IECC lists both general and specific requirements, often in different sections, that may apply to the same condition or situation. When two different code provisions apply to the same condition or situation, the more specific requirement applies to the design and construction, rather than the more general requirement, as in the other companion I-Codes. [Ref. C101.4]

EXISTING BUILDINGS

The term "existing buildings" refers to buildings constructed before the code's provisions took effect. The IECC does not require a legally constructed (permitted) existing building or building system to comply with the energy code provisions for new buildings. However, buildings constructed without a permit in a jurisdiction that administers building codes are not covered by this exemption. In addition, proposed work in any existing building that requires a permit must meet the IECC provisions, and is not exempt. [Ref. C101.4.1]

FIGURE 3-9 Historic building

FIGURE 3-10 Historic building designation

HISTORIC BUILDINGS

Certain existing buildings are currently listed, designated, or certified as historic or in the process of becoming certified as historic (Figures 3-9 and 3-10). Alterations or repairs to such buildings are exempt from the provisions of the IECC. Most communities have historic preservation programs, which can be valuable sources of information in determining whether a particular building is or could be certified as historic. The historic building exemption applies to mechanical, electrical, and plumbing systems as well as wall and roof insulation and window replacement. Alterations of historic commercial buildings are quite complex, as many unique issues are involved in maintaining the delicate systems in balance. **[Ref. C101.4.2]**

In 1986, the Empire State Building (Figure 3-11) was recognized as a National Historic Landmark. Construction of the iconic building began in 1930 and was completed in 1931, when there were no energy codes in place. The New York City Energy Conservation Code, an above-code program, is now the law of the land. **[Ref. C102.1.1]**

In 2010, a major rehabilitation project for the Empire State Building was announced, and certain energy-related retrofits were included in the plan. Major components of the energy upgrades include improvements to the windows, heating components and cooling system, and installation of lighting and ventilation controls. All 6,514 windows were refurbished on-site. The existing dual-pane glass was removed from the frame, cleaned, and reassembled with new seals. At the same time a suspended film was installed between the panes and the window unit was filled with inert gas. Note that the existing glass, sash, and frame of each window were reused to maintain the original materials and historic look. Inside, the radiators were removed and insulation barriers were installed behind the heating units on the exterior walls. Energy-efficient improvements were made

FIGURE 3-11 Empire State Building

to the chiller plant. Automatic sensors controlling light and ventilation were installed in public and individual tenant office spaces.

The building's current energy use is being measured, and a 38 percent improvement in energy efficiency is expected due to the retrofits. The work, although not required due to the historic building designation, is based on the energy-efficient principles, intent, and innovative techniques supported in the IECC.

WORK REQUIRING PERMITS

Most existing buildings are not considered historic, so any new work in these buildings—such as additions and renovations—must comply with the energy code provisions required for new construction. Parts of the building not affected by the new construction do not need to meet code provisions. This approach, "if you work on it, then it must comply," is consistent in all the I-Codes.

Although new work must comply with the code, there are specific exceptions—for example, projects or renovations that will not increase the building's energy use are exempt. However, installing a new window assembly in an existing window opening requires the window to meet the same requirements for solar heat gain coefficient (SHGC) and U-factor ratings as would be required for new construction. Typical energy-upgrade projects that include window replacement require permit application, plan review, permit issuance, and inspection.

Commercial tenant spaces that rearrange lighting only require a permit if more than half of the lighting fixtures in the space are replaced or the new lighting design increases energy use in the space. Compliance is usually confirmed with a lighting plan and fixture schedule. [Ref. C101.4.3]

CHANGE IN USE OR OCCUPANCY

Commercial tenant occupancy changes may trigger review of energy code compliance. Although residential apartment tenants come and go, the space is still an apartment. As commercial tenants come and go, the code requirements may change due to the new tenant's finish work for the building space. For example, the tenant finish work does not require energy code compliance if no alterations are made to the lighting systems, but if changes are made the new lighting system cannot use more power than is allowed for new construction. This may happen when a retail tenant moves into an existing office space and installs merchandise lighting. The new lighting displays may use more power than office lighting, but cannot use more than is allowed by the code for new construction. An example of a total and complete change of use and occupancy is a house converted into an office space or business (Figure 3-12). [Ref. C101.4.4]

Similar to the requirement for a change in occupancy is when any previously unconditioned space becomes heated or cooled. A warehouse or storage area that becomes an office or work area must comply with all

© International Code Council

FIGURE 3-12 Residential building converted into an office

the provisions of the code, which may require insulation changes as well as the replacement of doors and windows. [**Ref. C101.4.5**]

All commercial building uses are regulated by the provisions in this part of the code. In buildings not more than three stories tall that contain both commercial and residential spaces, the office and retail uses must comply with these commercial provisions. Residential spaces in a building not more than three stories high are regulated by the residential provisions of the code. Hotel or apartment buildings four stories or more, even though the entire use is residential, must meet the commercial energy provisions. [**Ref. C101.4.6**]

COMPLIANCE

Designers, builders, and code officials all must agree about the energy provisions applicable to a proposed project. All plans, specifications, and details—whether concerning new construction or the alteration or repair of an existing building—must comply with the minimum provisions of the code.

Some jurisdictions offer prescriptive worksheets that establish the requirements for building components and systems applicable in that location. This method is the easiest compliance path for simple commercial structures: Requirements for each building element are listed on the worksheets, and the plans are checked for compliance. The worksheets are specific to each climate zone and are often available from the state or local building department. [**Ref. C101.5.1**]

In addition, the U.S. Department of Energy offers ComCheck, a free and easy-to-use software program for verifying code compliance. The code official must approve the use of specific computer software such as ComCheck. In such programs, the user inputs building areas, efficiencies, and other specifications for the building envelope, mechanical systems, and interior and exterior lighting systems. The software then generates a compliance report for the approved plans and a customized field inspection checklist.

ComCheck compliance programs are available for editions of ASHRAE 90.1, the IECC, and specific state programs. The choices available in codes to match multiple jurisdictional adoptions make this a versatile compliance tool.

ALTERNATE METHODS AND MATERIALS

Because innovation is encouraged by the IECC, materials or methods of construction not specified in the code but meeting the intent of the applicable provision can be used if such are evaluated by the code official and approved before the building is permitted and inspected. The code official will consider the approval of the material or method based on the intent of the code provision that most closely describes the alternate. [Ref. C102.1]

Local, regional, and national energy-efficient programs that exceed the requirements of the IECC also can be considered by the code official to comply with the provisions of the energy code; however, the mandatory energy efficiency provisions in Chapter 4 of the IECC must be met. [Ref. C102.1.1]

Code Basics

Alternate materials and methods and above-code programs:

- The permit applicant is responsible for supplying information supporting the proposed alternate.
- The code official is responsible for evaluating the information to determine equivalency and compliance.
- The authority having jurisdiction to administer the building code may accept a program that exceeds the energy efficiency provisions of this code. •

SECTION 4

Administration and Enforcement

© istockphoto.com/mustafa deliormanli

PREPARING THE PLANS

The primary role of the permit applicant is to coordinate the skills of the design team to prepare a complete set of construction documents for the project, which will then be reviewed by a plans examiner for compliance with the *International Energy Conservation Code* (IECC). A more complete understanding of the specific provisions and requirements of the code by the design professionals and contractors will help the plans examiner to review the project for compliance and issue the permit.

In some cases, a set of building plans with details and schedules will describe a simple project sufficiently to demonstrate compliance with the code. However, for many commercial building projects, a complex set of documents—including mechanical, electrical, and plumbing plans—must be prepared by registered design professionals. It is important to know, in the early phase of design development, if the jurisdiction of the proposed building requires the plans to be prepared by a registered design professional. The code official is authorized to waive this requirement if it is determined that a registered design professional's stamp and seal is not required to confirm compliance with this code. [**Ref. C103.1**]

Elements of the building envelope, mechanical system drawings and schedules, service water heating information, and electrical and lighting system drawings and luminaire (lighting) schedules must be included to make a complete set of drawings for review of energy code compliance.

The prescriptive or performance compliance path will determine the detail required to demonstrate compliance with the code. The prescriptive provisions provide a simple path to the preparation of compliant building documents, as described next.

THE CONSTRUCTION DRAWINGS AND DOCUMENTS

The Thermal Envelope and Air Sealing

The plans must include building elevations depicting window, door, and skylight areas and wall sections showing the type of insulation and its thickness. This applies to a single-story retail building, a 4-story apartment building, and a 40-story office building. The thermal envelope must be continuous on every element of the exterior wall, roof, and fenestration detail. The design documents must have sufficient detail to confirm that all six sides of the building cube address insulation continuity. The thermal envelope must be completely described by the drawings.

These wall and roof sections will list the R-value of the insulation, and the window and door schedules will list the U-factors and solar heat gain coefficients (SHGCs) of the fenestration specified for the project. Wall components and framing must be identified separately. This is important, as the insulation and material requirements for wood- versus metal-framed walls are very different. Building plan elevations indicating window and door sizes and locations, and a roof plan showing skylights, must agree with the schedules, notes, and callouts. Caulking and sealing details for window and door frame installation must be included in the construction documents submitted for plan review. These notes and details are typical of how all exterior joints are to be sealed.

Air infiltration is the unintentional and uncontrolled movement of outside air into the building. The air usually moves through cracks and holes in the building envelope and doors used for egress. These same air-movement pathways allow for *air exfiltration*—inside air moving through the building envelope to the outside. The three primary causes of air leakage are wind, the stack effect, and mechanical equipment in the building. Wind creates pressure differences on opposite faces of the building. (Figure 4-1) The stack effect creates unequal pressures resulting from the density of the differences between heated interior air and exterior cold air or cooled interior air and warm exterior air. This is magnified by the height of the building. HVAC exhaust fans and systems, and supply fans and systems can also lead to pressure differences in a building. A tight building envelope helps control air movement and will help lower heating and cooling costs, make the building more comfortable to live and work in, and improve the performance of the HVAC system.

Code Basics

In contrast to *air infiltration, air barrier* is defined in the code as: "Material(s) assembled and joined together to provide a barrier to air leakage through the building envelope." •

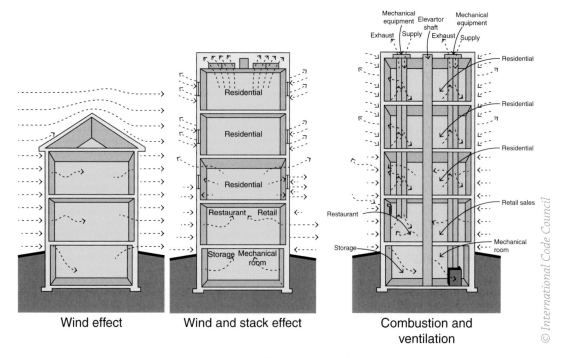

© International Code Council

FIGURE 4-1 Examples of air infiltration

You Should Know

Commercial building exterior wall assemblies often present very different challenges than those typical in residential construction. For example, the structural design requirements in a high-rise building include a strong *and* flexible frame design. The exterior cladding system may be made from glass, metal, stone, or a variety of other materials to create an architecturally pleasing face for the structure. The thermal barrier fits into wall sections between the interior wall finish and the exterior cladding.

Material choices for insulation must comply with the R-values or U-factors required by the energy code, and the flame-spread and smoke-development characteristics must comply with the applicable provisions in the building code. Section 720 of the *International Building Code* (IBC) contains the requirements for thermal insulation in walls, roofs, and around pipes. Section 1408 specifically addresses exterior insulation and finish system (EIFS) materials, construction, and inspection. The requirements for insulation installed above the roof deck are noted in Section 1508. Specific provisions for foam plastic material and installation are found in IBC Section 2603.

Fire-resistance-rated construction is an essential element of commercial building design and fire protection. The flame-spread restriction for materials helps control the growth of fire. Building occupants and rescue personnel must deal with smoke in a fire situation; lower limits of the smoke-development properties of insulation materials maximize time for occupants to exit the building and for firefighters to react to the situation. The fire-resistive properties of building materials and insulation must be met as required for the life-safety provisions in the IBC. The intent is that the registered design professional team will select insulation material and provide building design details that comply with both the IECC and the IBC. ●

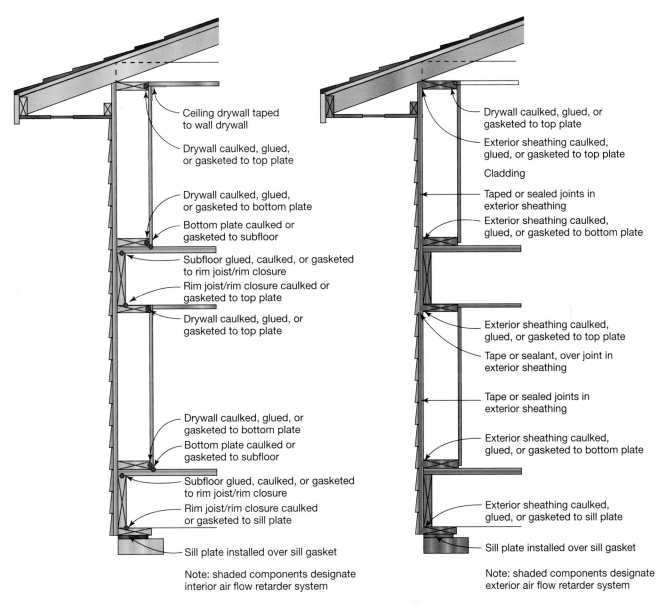

Ceiling drywall taped to wall drywall

Drywall caulked, glued, or gasketed to top plate

Drywall caulked, glued, or gasketed to bottom plate

Bottom plate caulked or gasketed to subfloor

Subfloor glued, caulked, or gasketed to rim joist/rim closure

Rim joist/rim closure caulked or gasketed to top plate

Drywall caulked, glued, or gasketed to top plate

Drywall caulked, glued, or gasketed to bottom plate

Bottom plate caulked or gasketed to subfloor

Subfloor glued, caulked, or gasketed to rim joist/rim closure

Rim joist/rim closure caulked or gasketed to sill plate

Sill plate installed over sill gasket

Note: shaded components designate interior air flow retarder system

Drywall caulked, glued, or gasketed to top plate

Exterior sheathing caulked, glued, or gasketed to top plate

Cladding

Taped or sealed joints in exterior sheathing

Exterior sheathing caulked, glued, or gasketed to bottom plate

Exterior sheathing caulked, glued, or gasketed to top plate

Tape or sealant, over joint in exterior sheathing

Tape or sealed joints in exterior sheathing

Exterior sheathing caulked, glued, or gasketed to bottom plate

Exterior sheathing caulked, glued, or gasketed to sill plate

Sill plate installed over sill gasket

Note: shaded components designate exterior air flow retarder system

FIGURE 4-2 Envelope air sealing

Air infiltration and exfiltration must be clearly addressed in the building plan documents. Again, the typical commercial building exterior wall is a complex assembly. All exterior joints, cracks, and holes are to be gasketed, caulked, weather-stripped, or sealed to create a continuous air-barrier layer. The air-sealing materials must be called out on the plans and compatible with the building materials in the joints and holes where they are applied. Details, notes, and sections in the building plans must address the continuity of the air sealing for the entire building envelope (Figure 4-2).

Mechanical System

Mechanical system design is particularly important in commercial construction. Most mechanical systems for commercial uses are sophisticated, complex, and designed by mechanical engineers (Figure 4-3).

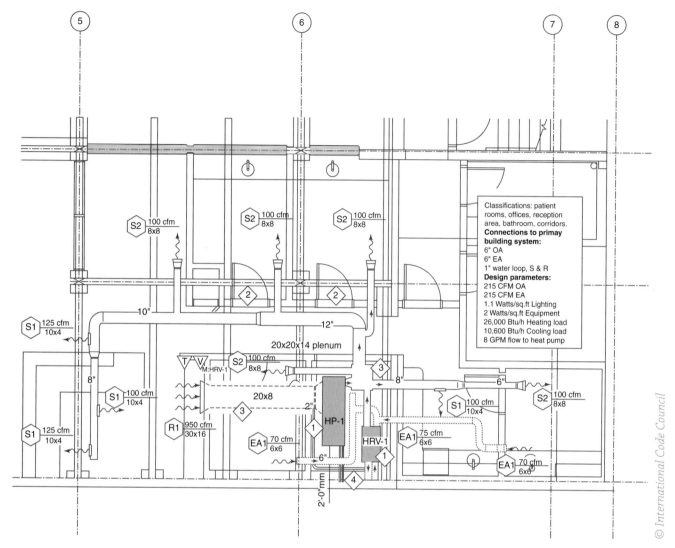

FIGURE 4-3 Mechanical plan

The documents require a statement describing design criteria and mechanical equipment schedules, including the types and efficiencies of the HVAC equipment and the systems controls.

The building ventilation system details must include fan motor horsepower and control system operation (Figure 4-4). Duct sealing is important to minimize uncontrolled conditioned air leakage in plenums and interior building spaces. Duct-sealing materials must be specified to demonstrate their compatibility with the duct materials. Air ducts and pipes carrying conditioned air or water are required to be insulated. The insulation minimizes heat loss or heat gain as the HVAC system delivers the conditioned air or heated or chilled water throughout the building.

Service Water Heating

Service water heating systems require separate design details and schedules for equipment efficiencies and control systems. The system piping layout, equipment location, and control operation information must be included in the design documents (Figure 4-5). Water heating equipment

			Mechanical Schedule						
Id	Description	Location	Manufacturer & Model	cfm	esp	Volt-age	Phase	HP/ watts	FLA
EA-1	Exhaust air grille, with integrated damper	Various	60 DAL-L , coordinate finish with Arch	–	–	–	–	–	–
HP-1	Horizontal water-to-air heat pump, condensate drain required	High in bathroom space	Water Furnace ND-H-026, with thermostat, 2" Merv 11 filter, hang with vibration isolation springs.	950	~ 0.2	230	1	373 watts	14.2
HRV-1	Heat recovery ventilator, condensate drain required	High in bathroom space	SHR 2004, EDF2 controller, hang with vibration isolation springs.	215	~ 0.3	115	1	150 watts	2.1
R-1	Return air grille	Various	60-F, coordinate finish with Arch	–	–	–	–	–	–
S-1	Spiral duct supply grille, with air scoop/damper	Various	SDGE-AS, air scoop, size for duct diameter (see plans), coordinate finish with Arch	–	–	–	–	–	–
S-2	Supply air grille, with integrated damper	Various	22DAL-ED-L, exposed duct connection, coordinate finish with Arch	–	–	–	–	–	–

FIGURE 4-4 Mechanical equipment schedule

size and efficiency schedules are required (Figure 4-6). Calculations may be included to demonstrate compliance with specific code provisions. Supplying the information the plan reviewer needs to determine compliance will speed the process of plan approval and permit issuance.

Lighting and Controls

Interior and exterior lighting systems and controls must be included in the design documents and must demonstrate compliance with the provisions of the IECC. The interior lighting system and switch-control strategy is usually described in the reflected ceiling plan (Figure 4-7). A lighting fixture schedule, including the wattage of each luminaire, is required (Figure 4-8). [Ref. C103.2]

PLAN REVIEW

The code official is charged with reviewing the construction documents for code compliance. [Ref. C103.3] It is not just the building plans but the entire suite of documents—including schedules, calculations, and specific product information—that is subject to review for code compliance. The final act in permit issuance is to stamp the approved document(s) "Reviewed for Code Compliance." One set of approved plans is kept in the building department office and an identical set is

Code Basics

A strategy for moisture and vapor control must be addressed in the building plan details. The requirements in IBC Section 1405.3 must be incorporated in the building plan wall section details. Exterior walls must be covered with materials approved as weather coverings and, in certain climate zones, provided with a vapor retarder. It is important to be familiar with and incorporate these provisions in the plan set. ●

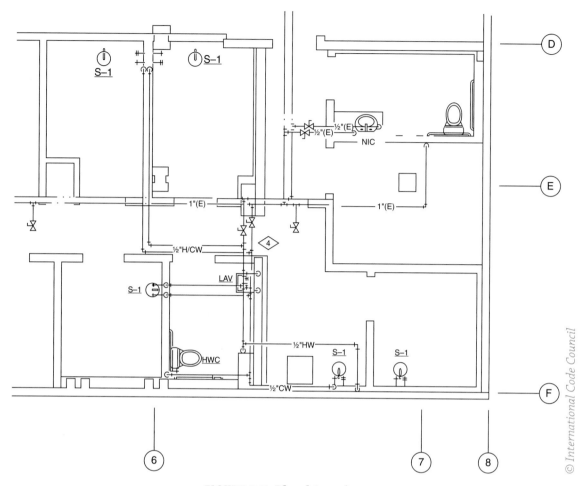

FIGURE 4-5 Plumbing plan

kept at the work site. This approved set of construction documents cannot be altered or changed without the code official's approval. [Ref. C103.3.1]

Officials will visit the work site for field inspections, and the construction must agree with the approved plans. Any changes to the approved plans must be resubmitted to the code official for further review, approved, and then the re-approved plans must be returned to the field before inspections are called for that portion of the work. [Ref. C103.4]

INSPECTIONS

Building construction work that requires a permit also requires inspection. Specific inspections are not listed in the IECC as they are in the other I-Codes. The periodic inspections required by the normal

Plumbing Fixture Schedule

Key	Description	Fittings/accessories	Manufacturer/catalog #
Lav	Square Wall-mounted Lavatory, ADA Compliant, Center Hole, White	Deck Mount Single Hole, Pop-up Drain, Offset Tailpiece, Insulated Drain Piping	Porcher 26020 - Center Hole Lavatory Delta 22c801
Hwc	Handicap Water Closet, Dual Flush Tank Type, Floor Mounted, Siphon Jet, Elongated Bowl, 1.0/1.65 Gallon, 16.5" Rim Height	Extra Heavy Duty Open Front, Solid Plastic Seat, Check Hinge	American Standard 2888.216 (Flo-wise) 5322.011 (Seat With Cover)
S-1	Single Bowl Undermount Sink, 20 Guage Stainless	Deck Mount Single Hole, Pop-up Drain	Kindred Qsu1816/8n Delta 22c801

Plumbing Fixture Pipe Sizes

Key	Hw Size	Cw Size
H/wc (flush tank)	-	1/2"
Lav	1/2"	1/2"
S-1	1/2"	1/2"

All pipe sizes as indicated except where noted.

FIGURE 4-6 Plumbing fixture schedule

construction process will incorporate many provisions of the energy code in the inspection schedule. For example, the foundation inspection requires perimeter and below-grade wall insulation to be in place before backfilling. During the framing or by special inspection for commercial construction, field staff will verify fenestration compliance and air sealing. The insulation inspection verifies that materials of the required R-values are in place and properly installed.

Mechanical, plumbing, and electrical inspections are required to verify compliance with the approved plans. The installations must be done safely and must comply with the requirements of the applicable codes as well as the approved plans. The schedules indicate the HVAC efficiency requirements; system controls; and duct size, location, and insulation. Service water heating inspections include verifying that equipment efficiencies match those documented on the plans, temperature controls are installed as approved, and pipe insulation is installed as required. Electrical and lighting inspections verify the number and type of fixtures installed according to the approved plans. Interior and exterior controls, switches, and timers must be installed according to the approved plans. **[Ref. C104.1]**

Specific energy code inspections are usually a matter of building department policy and procedure. A final inspection is the only required inspection mentioned in the IECC. **[Ref. C104.3]** This implies that all other field verifications for compliance blend with the community

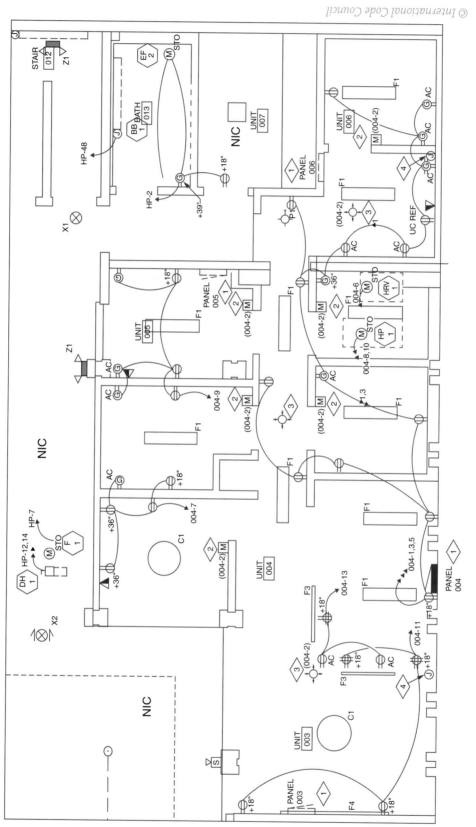

FIGURE 4-7 Electrical plan

Luminaire Schedule

Key	Lamp	Description	Ceil'g (Depth)	Manufacturer/#	Volt
C1	42w Gu-24 Cfl	Round Contemporary Chandelier, Coordinate Finish With Architect	Pendant, Verify Height W/ Architect	Lampa Satellite Chandelier	120
F1	(2)F32/t8 Spx35	Linear Fluorescent Pendant	Pendant, Verify Height W/ Architect	Metalumen Planar S2e	120
F3	Led	4' Straight & Narrow Led, Direct, White, Powder Coated Acrylic	Pendant, Verify Height W/ Architect	Cooper Lighting Neo-ray 22-d-p	120
F4	Led	8' Straight & Narrow Led, Direct, White, Powder Coated Acrylic	Pendant, Verify Height W/ Architect	Cooper Lighting Neo-ray 22-d-p	120
P1	75w Par 30	Line Voltage Direction Spotlight, White	Surface	Artemide Architectural Starship Surface	120

Section 2: Interior Lighting And Power Calculation

A Area Category	B Floor Area (Ft2)	C Allowed Watts/ft2	D Allowed Watts (B X C)
Healthcare-clinic	971	1	971
		Total Allowed Watts =	971

Area Category Exemption Qualifications

	Total Wattage		Total Pre-alt.	# Fixtures
Activity Area	Pre-alt.	Post-alt.	Fixtures	Repl./added

Section 3: Interior Lighting Fixture Schedule

A Fixture Id : Description/lamp/wattage Per Lamp/ballast	B Lamps/ Fixture	C #of Fixtures	D Fixture Watt.	E (C X D)
Healthcare-clinic (971 Sq.ft.)				
Linear Fluorescent 1: F1: Suspended Flourescent /48" T8 32w/electronic	2	10	64	640
Compact Fluorescent 1: C1: Satellite Chandelier/triple 4-pin 42w/electronic	1	2	42	84
Led 3: P1: Ceilint Mounted Monopoint/other	1	1	23	23
Led 1: F3: Suspended Led/other	1	2	45	90
Led 2: F4: Wall Mounted Led/other	2	1	90	90
		Total Proposed Watts =		927

FIGURE 4-8 Luminaire schedule

standards regarding the normal sequence of such inspections in the jurisdiction. Contractors and tradespeople are responsible for knowing the local rules regarding field inspection expectations.

FEES

Every jurisdiction that adopts building codes establishes a fee schedule. Fee amounts vary, so it is important to get specific information from the jurisdiction in which the proposed project is located. Any and all applicable fees must be paid before a specific permit is issued. Energy code compliance may require building, mechanical, plumbing, and electrical permits. Any work that needs a permit should not begin before the permit is issued. [Ref. C107.1] This action is subject to additional fees added to the permit fee. [Ref. C107.3] It is not a good idea to start work before a permit is issued because it may not comply with the applicable code provisions. This may be a costly mistake, taking more time, money, and materials than necessary.

ENFORCEMENT

A stop work order is the ultimate enforcement tool (Figure 4-9). The code official has the authority to stop work on any part of a project if the work regulated by the code is dangerous, unsafe, or not code compliant. [Ref. C108.1] The stop work order must be in writing, is issued to the owner or person doing the work, and must be specific to the noncompliant work ordered to be stopped. [Ref. C108.2] If the stop work order is ignored, the code official may fine the person responsible, an action that generally involves a citation in a municipal court. The amount of the fine is determined by each jurisdiction and written into the code at its adoption. [Ref. C108.4]

BOARD OF APPEALS

The board of appeals exists to allow a means to appeal decisions made by the building official. Members of boards of appeal are usually civic-minded architects, engineers, insulators, lighting designers, or mechanical, plumbing, or electrical contractors. [Ref. C109.3] The board of appeals is created to hear and decide whether the appellant or code official's action is consistent with the intent of the code. For example, it may be the case that the IECC does not apply to a specific project condition, or that an equally good or better alternative approach to a proposed construction method is rejected by the code official. It is the broad understanding of the energy code that qualifies board members to hear such appeals. [Ref. C109.3]

Note that it is the *intent* of a certain code provision that generates the basis for an appeal. An appeal may be heard when an appellant—perhaps an owner, designer, or tradesperson—does not agree with the code official's decision or order. The appeal may concern a design or field inspection decision that the appellant believes is incorrect based on the true intent of the code provision or rule as applied by the code official. [Ref. C109.1] The board hears the appeal, makes a decision, and records the findings in writing. In no case may the board waive any requirements of the code. [Ref. C109.2]

REGIONAL BUILDING DEPARTMENT
☐ CORRECTION NOTICE
☐ STOP WORK ORDER

Job Located at _____

I have this day inspected this structure and these premises
and have found the following violations of City, County and/or
State laws governing same:

_____

_____

_____

_____

_____

_____

_____

_____

_____

_____

Respond By _____ Photos Taken : Yes ☐ No ☐
 You are hereby notified that no more work may be done upon the premises until the above violations are corrected. If you do not communicate with this office by the above date, this matter will be referred to the appropriate authorities for enforcement. Failure to correct the violations may subject you to a civil suit for an injunction, or a fine, or both; or to misdemeanor criminal prosecution, which upon conviction may carry a sentence of fine or imprisonment, or both.

_____   _____
Date   Inspector for Building Department

Building Department Phone:
DO NOT REMOVE THIS TAG

FIGURE 4-9 Typical red tag form

Code Basics

Terms that are not defined in this code (2012 IECC) but are defined in the International Building Code, International Fire Code, International Fuel Gas Code, International Mechanical Code, International Plumbing Code or the International Residential Code shall have the meanings ascribed to them in those codes. ●

DEFINED TERMS

The 85 specific words and terms in Section 2 of the IECC define what code users agree to understand as the meaning of such terms when written and appearing in the code provisions. **[Ref. C201]** For example, the words *accessible, readily accessible, building, listed, repair, storefront,* and *zone* all are defined. Any defined word or term appears in *italics* in the code text. In reading, interpreting, and applying code provisions, it is essential to use the common code vocabulary. **[Ref. C202]** This is the basis for civil discussion and deeper understanding of the energy code provisions.

PART

III

SPECIFIC REQUIREMENTS FOR COMMERCIAL BUILDINGS

© Orhan Cam/www.Shutterstock.com

Requirements for Commercial Buildings

© istockphoto.com/mustafa deliormanli

Design criteria must be established to guide engineers and builders to regional and specific climate and rainfall data most appropriate to the area where the building will be constructed. The climate zone map references every county in the continental United States and Hawaii and the territories of Guam, Puerto Rico, and the Virgin Islands. In fact, design criteria for any location in the world can be deduced from the international climate zone definitions. Overall building performance relies on materials not only having the proper minimum efficiency but also being properly installed.

CLIMATE ZONES

The climate zone map is a simple and easy-to-understand graphic depiction of historical weather data (Figure 5-1). The colorful visual map is supplemented by Table C301.1 in the *International Energy Conservation Code* (IECC), which lists climate zones, moisture regimes, and warm-humid designations by state, county, and territory. This section also defines climate types (moisture regimes) and thermal criteria so that any place in the world with reliable weather data can be assigned a climate zone and climate type. [Ref. C301.1] This information is the starting point in energy-efficient building design. All of the prescriptive building envelope provisions are based on weather data for the building's proposed location. The climate zone map is recognized in the development community as the important base document to fulfill the intent of the code to "regulate the design and construction of buildings for the effective use and conservation of energy over the useful life of each building."

Heating Degree Days (HDD) and Cooling Degree Days (CDD) climate data define the separation criteria of the zones. Both are reliable indicators for predicable energy use and determine insulation and fenestration requirements and the performance of heating and cooling equipment. A basic knowledge of these terms is helpful in understanding the need for the prescriptive provisions of the code. The greater the HDD, the better the building must be designed to perform in cold weather. The greater the CDD, the better the building must be designed to perform in hot weather. Insulation R-values are relaxed in warmer climates, and values for solar heat gain coefficient (SHGC) are also relaxed in colder climates.

You Should Know

Warm humid counties are identified in Table C301.1 by an asterisk. ●

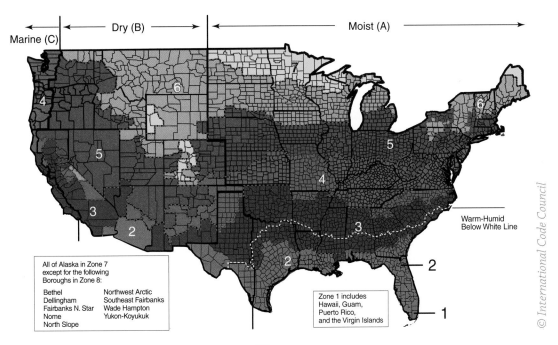

FIGURE 5-1 Climate zone map

FIGURE 5-2 Marine climate conditions

FIGURE 5-3 Moist climate conditions

FIGURE 5-4 Dry climate conditions

Building envelope components are adjusted to fit the conditions of the building's location.

The climate zone map also recognizes the moisture-regime designations of marine, dry, and moist conditions, as well as warm-humid counties (Figures 5-2, 5-3, and 5-4). The designations are based on the combination and duration of seasonal temperatures and precipitation. Provisions for energy recovery ventilation and economizer mechanical equipment are determined by the climate and moisture-regime designations.

Code Basics

Section 1204.1 of the *International Building Code* (IBC) and Section R303.9 of the *International Residential Code* (IRC) establish the minimum interior comfort temperature for heating. •

DESIGN CONDITIONS

Interior design temperatures are established in the code to properly size heating and cooling mechanical equipment. The maximum design temperature for heating equipment is 72 degrees. The minimum design temperature for cooling is 75 degrees. **[Ref. C302.1]** It is very important to note that these criteria provide essential input for mechanical equipment design; however, they do not establish the interior comfort or livability provisions addressed in other building codes.

INSULATION MATERIALS

Insulation materials used in the building thermal envelope must somehow be marked and identified to confirm compliance with the approved plans. [Ref. C303.1] Each piece of batt insulation 12 or more inches in width must be marked by the manufacturer (Figure 5-5).

If the product is not stamped at the factory, the installer must provide a certificate listing the insulation values and information for each area in which the unmarked insulation is installed in the building thermal envelope. Blown or sprayed fiberglass and cellulose wall insulation require certification from the installer specifying thickness and R-value details. The certificate for sprayed polyurethane foam (SPF) requires the R-value of the installed thickness of the foam (Figure 5-6). In all cases, the installer must sign, date, and post the certification at the job site. [Ref. C303.1.1]

Fiberglass and cellulose spray insulation in roof/ceiling thermal assemblies is measured with markers indicating each 1-inch thickness of installed insulation (Figure 5-7). The markers are attached to the trusses or joists and face the attic opening. [Ref. C303.1.1.1] This arrangement makes it easy for the contractor and building inspector to verify compliance without crawling through the attic with a tape measure.

FIGURE 5-5 Unfaced insulation marked on bag

© International Code Council

You Should Know

For fiberglass or cellulose blown-in or sprayed insulation the certification must include 6 characteristics, one of which is installed density. •

FENESTRATION PRODUCTS

The National Fenestration Rating Council (NFRC) develops testing criteria for whole-assembly performance. The window, door, or skylight glass, seals, frame, and latching mechanisms are tested to produce the U-factor energy performance rating. The results are printed on a label attached to

© International Code Council

FIGURE 5-6 Measuring spray foam

© International Code Council

FIGURE 5-7 Depth marker for blown-in attic insulation

each product. The label indicates the qualities of the fenestration, allowing the builder and inspector to confirm compliance with the approved plans and schedules. Not every window, door, or skylight manufacturer participates in the NFRC testing program, and some products will appear on a job site without a factory-affixed label. In that case, the default tables in Section C303 of the IECC determine the assumed U-factor for the fenestration. Each fenestration product is assigned a value by the label or by the default table. **[Ref. C303.1.3]**

EXPOSED FOUNDATION INSULATION

The code requires insulation on the exterior foundation to be protected from physical damage and direct sunlight. Basement and crawlspace walls and slab-edge insulation must be covered with a rigid, opaque, weather-resistant material that covers the exposed material above grade and extends at least 6 inches below grade (Figure 5-8). **[Ref. C303.2.1]** This protection contributes to maintaining the thermal design performance of the installed insulation.

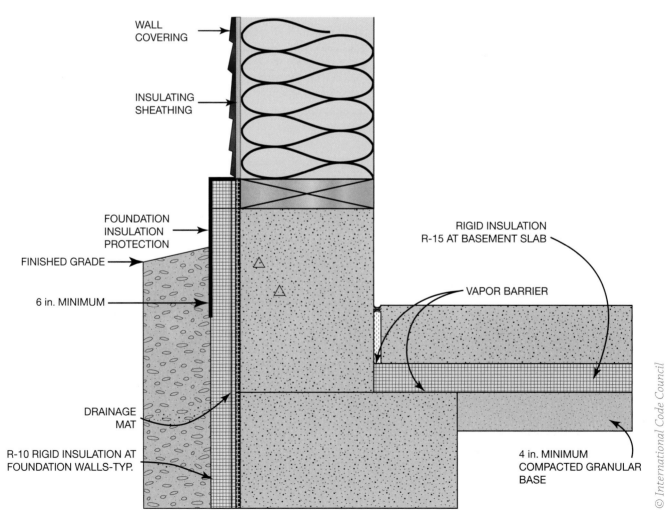

WALL COVERING

INSULATING SHEATHING

FOUNDATION INSULATION PROTECTION

FINISHED GRADE

6 in. MINIMUM

DRAINAGE MAT

R-10 RIGID INSULATION AT FOUNDATION WALLS-TYP.

RIGID INSULATION R-15 AT BASEMENT SLAB

VAPOR BARRIER

4 in. MINIMUM COMPACTED GRANULAR BASE

© International Code Council

FIGURE 5-8 Exposed foundation insulation protection

Efficiency Requirements

© istockphoto.com/mustafa deliormanli

The energy-efficient provisions for the building envelope, mechanical systems, service water heating, and electrical power and lighting systems of commercial buildings are contained in Chapter 4 of the *International Energy Conservation Code* (IECC). Section C402 provides prescriptive requirements for building envelope code compliance. This section offers a simple guide in the form of R-value and U-factor tables arranged by building envelope components, and answers the question "What do I have to do to comply?" The building envelope must be designed to meet the requirements of Table C402.2 or Table C402.1.2. Mandatory provisions for air leakage and air barriers, mechanical systems, pipe insulation and water heating, and lighting systems are included.

Section C406 of the IECC requires one of three choices of additional efficiency options. For code compliance, the designer or building owner must include either higher HVAC performance or efficient lighting systems beyond the mandatory requirements, or install an on-site renewable energy supply to satisfy this provision. **[Ref. C401]**

Commissioning certain building mechanical and electrical power and lighting systems is required in C408. The commissioning agent will verify that the performance of these systems meets the criteria specified in the commissioning plan and the completion requirements of the design documents.

Typical construction techniques and materials used in high-rise office and apartment buildings and commercial businesses differ from those in typical low-rise residential buildings. It should be noted that most jurisdictions and state statutes require commercial architectural and engineering plan documents to be stamped and sealed by a licensed and registered professional. The following overview is limited to highlight the basic concepts of the commercial energy design provisions, given the complexity of commercial building design.

BUILDING ENVELOPE WALLS AND ROOFS

The building thermal envelope separates the inside environment from the outside environment that is the daily and seasonal exterior temperature swings and moisture events. The typical elements of the thermal envelope include roofs and above- and below-grade walls and floors, as listed in the prescriptive tables C402.1.2 and C402.2. The key to establishing the boundaries of the thermal envelope is found in the definition of "conditioned space." For the purpose of the IECC, *conditioned space* is "an area or room within a building being heated or cooled, containing uninsulated ducts, or with a fixed opening directly into an adjacent conditioned space." It is important that the project designer defines the limits of the building thermal envelope in the plans and demonstrates compliance with the mandatory and applicable requirements of Chapter 4 of the IECC.

As noted, the code offers two choices for prescriptive compliance: R-value and U-value ratings. **[Ref. C402.1]** Table C402.2 lists R-value requirements for individual components of a typical commercial building envelope and the required performance level by climate zones. The values are also dependent upon the occupancy and are designated as either "Group R" or "all other" (Figure 6-1).

Not every building design will use every component, and it is important to be familiar with the definitions and descriptions of these

Table C402.2	
Climate Zone 5	**Commercial Building**
Insulation entirely above deck	R-25ci
Mass wall-above grade	R-11.4ci
Unheated slabs	R-10 for 24" below

© International Code Council

FIGURE 6-1 R-value prescriptive requirements

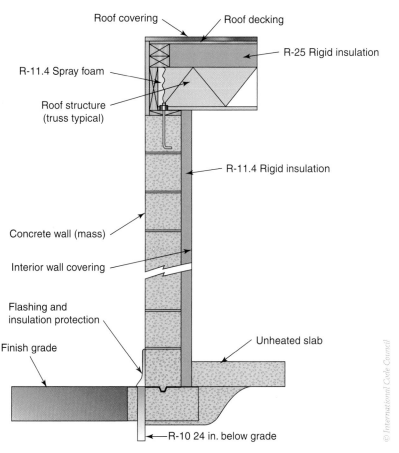

FIGURE 6-2 R-value insulation requirements for commercial building in climate zone 5

components. Only when building components are clearly identified in the drawings can the minimum R-values be specified for the roof, wall, and floor types listed in the table (Figure 6-2). **[Ref C402.1.1].**

Table C402.1.2 is the U-value alternative to Table C402.2. The applicable building envelope component is evaluated by an acceptable method, and the supporting calculations are submitted with the plan documents for review and approval (Figure 6-3). **[Ref C402.1.2].**

WINDOWS, SKYLIGHTS, AND DOORS

Table C402.3 of the IECC lists the prescriptive compliance requirements for commercial fenestration. The maximum window area allowed is 30 percent of the gross above-grade wall area. Skylight area is limited to not more than 3 percent. Certain exceptions are granted in areas with daylighting controls that allow the areas to be increased to 40 percent of above-grade walls and 5 percent of the roof area. The solar heat gain coefficient (SHGC) and projection factor are considered in establishing the requirements in this table. SHGC is an important consideration in energy efficiency and must be balanced with visible light transmittance in glazing design. The projection factor is a building element design strategy to provide shading to offset a portion of the heat gain through

ENVELOPE COMPLIANCE CERTIFICATE

2012 IECC

Section 1: Project Information

Project Type: **New Construction**
Project Title:

Construction Site: Owner/Agent: Designer/Contractor:

Section 2: General Information

Building Location (for weather data):	**Mountain Town, Colorado**
Climate Zone:	**7**
Building Type for Envelope Requirements:	**Non-Residential**
Vertical Glazing/Wall Area Pct.:	**1 percent**

Activity Type(s)	Floor Area
Common Areas (Office)	5049
Basement Level (Retail)	2759
1st-Floor (Retail)	3044
2nd-Floor Office (Office)	4040

Section 3: Requirements Checklist

Envelope PASSES: Design 5 percent better than code.

Climate-Specific Requirements:

Component Name/Description	Gross Area or Perimeter	Cavity R-Value	Cont. R-Value	Proposed U-Factor	Budget U-Factor(a)
Basement Wall 1: Solid Concrete:12" Thickness, Medium Density, Furring: Metal, Wall Ht 13.2, Depth B.G. 13.2	64320	23.0	4.0	0.080	0.108
Exterior Wall 1st Level: Steel-Framed, 16" o.c.	66937	40.0	1.5	0.057	0.064
Window 1: Wood Frame: Double Pane with Low-E, Clear, SHGC 0.60, PF 0.44	414	–	–	0.250	0.350
Window 2: Wood Frame: Double Pane with Low-E, Clear, SHGC 0.60, PF1.43	78	–	–	0.250	0.350
Door 1: Glass (> 50 percent glazing): Nonmetal Frame, Entrance Door, SHGC 0.60, PF 0.44	66	–	–	0.250	0.350
Door 2: Insulated Metal, Non-Swinging	21	–	–	0.300	0.500
Door - garage door: Insulated Metal, Non-Swinging	144	–	–	0.500	0.500
Door 4: Insulated Metal, Non-Swinging	21	–	–	0.300	0.500
Exterior Wall 2nd Level: Steel-Framed, 16" o.c.	56385	40.0	1.5	0.057	0.064
Window 3: Wood Frame: Double Pane with Low-E, Clear, SHGC 0.60, PF 0.02	437	–	–	0.250	0.350
Window 4: Wood Frame: Double Pane with Low-E, Clear, SHGC 0.60	405	–	–	0.250	0.350
Door 5: Insulated Metal, Non-Swinging	21	–	–	0.300	0.500

FIGURE 6-3 U-value prescriptive compliance certificate

the fenestration. Dynamic glazing is recognized in the IECC and must be considered separately for compliance with the U-factor and SHGC requirements. [Ref C402.3.3.5]

PERFORMANCE COMPLIANCE

The *total building performance* provisions of section C407 of the IECC offer a third path for energy code compliance. This method uses computer software to predict the annual energy cost of the proposed design as compared to the standard design of the building components listed in Table C407.5.1(1). If the proposed building's annual energy cost is less than or equal to that of the standard design, the building complies with the code's goal of effective energy use. Heating systems, cooling systems, service water heating systems, lighting power, receptacle loads, and process loads must be included in the calculations to determine the total building performance. [Ref C407.1]

This performance method allows trade-offs between the building envelope, HVAC system, and lightings systems to achieve an overall building energy conservation result equal to or better than that of the other code compliance paths. These individual component requirements in the modeling may be less than what is typically specified in the energy code tables. No matter the component performance details of the building envelope, HVAC system, and lighting systems, certain specific mandatory requirements must be met. The compliance report must be submitted to the code official. [Ref C407.4.1]

You Should Know

The following factors are often used to perform a Total Building Performance:

- Orientation
- Fenestration Area
- Size of Overhangs
- Fenestration U-Factor
- Visible Light Transmittance and Daylighting Control Fraction (if automatic daylighting controls are used)
- U-factors of building envelope components
- Heat Capacity
- Insulation Position (for mass walls only) ●

Controlling Air Leakage

© International Code Council

Uncontrolled outdoor-to-indoor air exchange through the building envelope is known as *infiltration*. The opposite air movement, indoor-to-outdoor exchange, is *exfiltration*. Either type of air exchange affects energy costs. Conditioned air, heated or cooled, lost through exfiltration must be replaced by air from the outdoors. The building air-handling systems then use additional energy to bring the outdoor air to the desired indoor temperature. By the same process, hot or cold air that infiltrates the building places an increased burden on the system because the air leaking into the building must be conditioned and brought to the desired indoor temperature.

Air exchange is controlled through intentional ventilation. This may be as simple as opening a window or, as in most commercial buildings, through the HVAC system. Excess air exchange, controlled or uncontrolled, is contrary to the intent of the IECC. [Ref. C101.3]

THE AIR-BARRIER REQUIREMENTS

The air-leakage provisions of the IECC are mandatory, and specific sections address air barriers and their construction, compliance requirements and options, penetrations, fenestration, air intakes and exhaust, loading dock seals, and vestibules. [Ref. C402.4] The requirements cover provisions for a continuous air barrier to be designed, detailed, and specified in the building plan set of documents. *Continuous air barrier* is defined in the IECC: "A combination of materials and assemblies that restrict or prevent the passage of air through the building thermal envelope." [Ref. C202] The air barrier may be inside or outside of the thermal envelope, and in all cases must be continuous. An easy test to check for compliance is to trace the air-barrier location in the building sections and details on the construction documents. If this can be done without lifting the pencil off the paper, the air barrier is continuous.

Air-barrier materials cover and seal the joints and assemblies in the building thermal envelope. A single material or a combination of materials is allowed to satisfy the requirement. The materials must be securely installed along the entire length of the joint in the thermal envelope and perform during the pressure changes in the building.

Compliance

Three methods are available to the designer and builder to comply with the air-barrier provisions for the solid portions of the building envelope: (1) compliance with a list of 15 prescriptive materials with a specific air permeability, (2) acceptance of assemblies with a specified average air-leakage rate, or (3) performance of a test of the completed building to verify the actual air-leakage rate. [Ref. C402.4.1.2]

In the first method, 15 common building materials are named specifically (Table 7-1). When any of these materials is installed according to the manufacturer's installation instructions and the joints are properly sealed, this requirement is met. Any material not on the list that meets the specified air-leakage rate when tested using the ASTM E 2178 test for air permanence also complies. [Ref. C402.4.1.2.1]

The second method relies on the performance of the assemblies versus the individual materials that the first option addressed. This method also includes two assemblies that are deemed to meet the required limits: the specific thicknesses of the parge, stucco, or plaster; or treatments to concrete masonry walls. The referenced tests apply to curtain walls, air-barrier systems, and large window assemblies. [Ref. C402.4.1.2.2] Pressure testing of the completed building is the third method to demonstrate air-barrier or air-leakage compliance. [Ref. C402.4.1.2.3]

You Should Know

ASTM E 2178 is the testing standard to measure air permeance of air barrier materials. Liquid applied membranes, mechanically fastened commercial building wraps and rigid panel materials are common air barrier materials in commercial construction. The test results of the specific materials are reported and must comply with the IECC. ●

TABLE 7-1 Common building materials that comply with the air leakage requirements

Air-Barrier Material (Prescriptive)	Thickness (minimum)
Plywood	3/8 in.
Oriented strand board	3/8 in.
Extruded polystyrene insulation board	1/2 in.
Foil-faced urethane insulation board	1/2 in.
Closed cell spray foam minimum density of 1.5 pcf	1-1/2 in.
Open cell spray foam density between 0.4 and 1.5 pcf	4.5 in.
Exterior gypsum sheathing or interior gypsum board	1/2 in.
Cement board	1/2 in.
Built-up roofing membrane	Thickness not applicable
Modified bituminous roof membrane	Thickness not applicable
Fully adhered single-ply roof membrane	Thickness not applicable
A Portland cement/sand parge, stucco, or gypsum plaster	5/8 in.
Cast-in-place and precast concrete	Thickness not applicable
Sheet metal or aluminum	Thickness not applicable
Fully grouted concrete block masonry	Thickness not applicable

© International Code Council

Penetrations

Penetrations in the air barrier must be sealed to maintain continuity. Electrical boxes, mechanical vents, and hose bibs are common penetrations that must be gasketed or caulked (Figure 7-1). Joints between materials create a break in the air barrier and must be caulked, taped or otherwise sealed. The sealing and air-barrier materials must be compatible and installed according to the manufacturer's installation instructions. [Ref. C402.4.2]

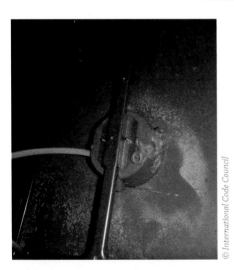

© International Code Council

FIGURE 7-1 Penetrations in the air barrier must be sealed

INFILTRATION RATES

Fenestration components are listed in Table C402.4.3 of the IECC and grouped by the applicable test procedure. Each fenestration component fills a hole in the opaque (solid) exterior building wall thermal envelope and air barrier. It is important to identify the windows, skylights, and most doors on the plans and schedules. Field inspection will confirm that the units are listed, labeled, and installed according to the manufacturer's instructions. Service-access openings to shafts and chutes and doors to stairways and elevator lobbies from conditioned space must be listed and gasketed, weather-stripped, or sealed.

OUTDOOR AIR OPENINGS

Mechanical ventilation systems bring in fresh air to the building interior and exhaust stale air to the outdoors. The openings must penetrate the air barrier to get the job done. Uncontrolled air flow through the outdoor air vents and in stairway enclosures and elevator shaft vents is accelerated by pressure imbalances between the building interior and exterior. Motorized dampers are required in these openings to control the air leakage through the air barrier (Figure 7-2). The dampers must be listed and tested.

Gravity dampers comply in buildings less than three stories above grade and for ventilation air intake, exhaust, and relief dampers in any building in climate zones 1, 2, and 3. **[Ref. C402.4.5.2 exceptions 1.2 and 1.3]** These particular exceptions are a good reminder to read and reread the code provisions completely, as they are an example of an opportunity to save significant time and money and still comply with the applicable energy code provisions for dampers.

© *International Code Council*

FIGURE 7-2 Motorized dampers control the air leakage through the air barrier

LOADING DOCKS

Most commercial buildings require loading docks to deliver food, dry goods, and supplies. The exterior

FIGURE 7-3 Compliant loading dock seal

FIGURE 7-4 Non-compliant loading dock

FIGURE 7-5 Revolving doors do not require vestibules

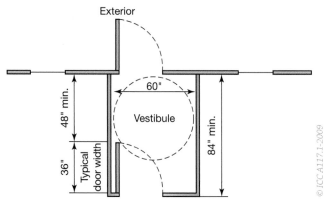

FIGURE 7-6 Typical minimum dimensions for a compliant vestibule

of the loading dock door perimeter must have a weatherseal installed to minimize the air flow (Figure 7-3). The provision applies to all loading dock doors in all climate zones. Loading dock doors are pictured in Figure 7-4. **[Ref. C402.4.6]**

VESTIBULES

The common architectural meaning for the term *vestibule* is "a passage, hall, or antechamber between the outer door and the interior parts of a house or building." This transition space, from outdoors to indoors, is a break in the air barrier and a direct path of infiltration every time the door is used. Vestibule provisions include self-closing doors and designs that allow one door behind you before opening the next. Although revolving doors do not require vestibules, any other entrance door adjacent to the revolving door does require a vestibule. Having a revolving door at the building entrance does not satisfy the intent of the vestibule requirement for these other doors (Figure 7-5). **[Ref. C402.4.7]**

In most buildings, vestibules designed to meet the energy code provisions result in significant energy savings. The IECC recognizes that some factors—such as the size of a compliant design, the use of certain exterior doors, and locations in warm climates—may create exemptions from the vestibule requirements. Practical exceptions include exterior doors in climate zones 1 and 2, doors that open into a space less than 3,000 square feet in area, and doors intended solely for employee use.

Accessibility must also be considered when designing the distance between doors. The building code and accessibility laws require certain minimum sizes for vestibules (Figure 7-6).

Mechanical Systems

© International Code Council

Mechanical system equipment produces and circulates tempered and treated air and hydronic fluids to make people comfortable inside the building. The energy efficiency of the HVAC system depends on it operating only when needed, being off when it is not needed, and using no more energy than necessary while it is operating. The mandatory provisions in the *International Energy Conservation Code* (IECC) **[Ref. C403.2]** to accomplish this include criteria for sizing the HVAC system and equipment, minimum equipment efficiency, system operational controls, ventilation volume controls, energy recovery ventilation (ERV) systems, air duct insulation and sealing, piping insulation, mechanical system commissioning, and controlling heat outside a building. Provisions specific to simple systems **[Ref. C403.3]** and complex systems **[Ref. C403.4]** are clearly divided in the separate code requirements. **[Ref. C403.1]** It is important to correctly classify the system as simple or complex to determine the specific code provisions in addition to the mandatory requirements for all HVAC systems.

THE MANDATORY REQUIREMENTS

HVAC Load Calculation

Accurate load calculations are essential and the first step in mechanical system design. ANSI/ASHRAE/ACCA Standard 183 establishes the calculation method, and any approved equivalent must follow the established procedures. ACCA *Manual N* is generally viewed as an approved equivalent.

It is important to note that the indoor design temperature input is limited to 72°F for heating and 75°F for cooling as required by C302. Load calculations must be submitted with the mechanical drawings and specifications.

Equipment Sizing

The design load calculation provides the information required to determine the right size of HVAC equipment for the building. The output rating of the heating and cooling equipment listed in the mechanical schedule must not be greater than the calculated heating and cooling loads [Ref. C403.2.2]. Undersized or oversized HVAC equipment is inefficient because the run cycles are either too long (undersized) or too short (oversized).

Equipment Efficiency

Heating and cooling equipment must meet the minimum efficiency requirements in Tables C403.2.3(1–8) of the IECC. The tables are grouped to help the user locate equipment type and equipment size. The detailed descriptions and size choices offered in the tables guide the designer and plans examiner in determining the minimum efficiency requirements specific to the equipment in the drawings and schedules (see Table 8-1).

The efficiency of air-conditioning equipment is rated by several metrics established in the referenced test procedure. The Air-Conditioning, Heating, and Refrigeration Institute (AHRI) offers a simple and understandable basis for establishing the standards used to compare this type of equipment. The seasonal energy efficiency ratio (SEER) is the ratio of the seasonal cooling output divided by the total electric energy input. The energy efficiency rating (EER) is the cooling output at a particular operating point divided by the electrical power input. The integrated energy efficient ratio (IEER) measures the part-load performance of the unit.[1]

Heating equipment efficiency is rated similarly to cooling equipment, using test metrics and standards. The U.S. Department of Energy explains annual fuel utilization efficiency (AFUE) as "the ratio of heat

Code Basics

The code specifies the design temperatures so the equipment can be properly sized. Design temperatures are based on climate data within the larger climate zones referenced in the IECC. The information allows the HVAC to be properly sized allowing efficient operation. Equipment designed to operate over the design temperature for heating or below the design temperature for cooling defeats the energy efficient operation and design intent of the IECC. ●

---

[1]ANSI/AHRI Standard 21/240 with Addenda 1 and 2 (formerly ARI Standard 210/240) 2008 *Standard for Performance Rating of Unitary Air-Conditioning & Air-Source Heat Pump Equipment.*

TABLE 8-1 Heating and Cooling Minimum Efficiency Requirements

Table C403.2.3(2) Minimum Efficiency Requirements: Electrically Operated Unitary and Applied Heat Pumps					
Equipment Type	Size Category	Heating Section Type	Subcategory or Rating Condition	Minimum Efficiency	Test Procedure
Air cooled (cooling mode)	< 65,000 Btu/h	All	Split System	13.0 SEER	AHRI 210/240
			Single Packaged	13.0 SEER	
Through-the-wall, air cooled	< 30,000 Btu/h	All	Split System	13.0 SEER	
			Single Packaged	13.0 SEER	
Single-duct high-velocity air cooled	< 65,000 Btu/h	All	Split System	10.0 SEER	
Air cooled (cooling mode)	≥ 65,000 Btu/h and < 135,000 Btu/h	Electric Resistance (or None)	Split System and Single Package	11.0 EER 11.2 IEER	AHRI 340/360
		All other	Split System and Single Package	10.8 EER 11.0 IEER	
	≥ 135,000 Btu/h and < 240,000 Btu/h	Electric Resistance (or None)	Split System and Single Package	10.6 EER 10.7 IEER	
		All other	Split System and Single Package	10.4 EER 10.5 IEER	
	≥ 240,000 Btu/h	Electric Resistance (or None)	Split System and Single Package	9.5 EER 9.6 IEER	
		All other	Split System and Single Package	9.3 EER 9.4 IEER	

output of the furnace or boiler compared to the total energy consumed by the furnace or boiler."[2] The coefficient of performance (COP) is the ratio of change in units of heat or cooling energy output by the equipment divided by the energy used for the output (Figure 8-1). A heat pump COP of 3.3 means 3.3 units of heat energy, expressed in kilowatts (3.3 kWh), are provided for every 1 unit of energy (1 kWh) used.

Equipment System Control

Each heating and cooling zone must be independently controlled by a programmable thermostat. A *zone* is defined as "A space or group of spaces within a building with heating or cooling requirements that are sufficiently similar so that desired conditions can be maintained throughout using a single control device." **[Ref. C202]** The control device must be capable of setting back the HVAC temperature demand

---

[2]U.S. Department of Energy, *Energy Efficiency & Renewable Energy*, "Understanding the Efficiency Rating of Furnaces and Boilers."

FIGURE 8-1 Output rating

when the building is not in use on weekdays, weekends, and holidays, and must be capable of starting the system to make the building comfortable before the building is occupied. **[Ref. C403.2.4.3]**

Motorized shutoff dampers are generally required on all outdoor air supply and exhaust ducts. The dampers limit air infiltration and exfiltration when the HVAC systems are not operating. The code provides exceptions allowing nonmotorized/gravity dampers to be used on certain buildings or applications. **[Ref. C403.2.4.4]**

Snow and Ice Melt System Controls

Snow and ice melt systems linked to the building's energy system, either hydronic (Figure 8-2) or electric, must have automatic controls. The controls monitor temperature and precipitation. When the pavement and air temperatures are safely above the point beyond which snow will not stick and ice will not accumulate, the system energy use is greatly reduced. The controls allow the system to idle in the off mode. This lessens the thermal shock effect and prolongs the functional life of the pavement slab. A manual control (Figure 8-3) is allowed so that the system can be shut off completely during warm weather. **[Ref. C403.2.4.5]**

You Should Know

In the 2012 IECC all HVAC systems are required to be capable of automatically adjusting the daily start time to bring each space to the desired occupied temperature immediately prior to scheduled occupancy. This is in addition to the manual override with timer or occupancy sensor that is required by Section C403.4.3.2. ●

FIGURE 8-2 Snow and ice melt systems

VENTILATION

The *International Mechanical Code* (IMC) requires ventilation in all buildings. Natural or mechanical ventilation is acceptable as long the minimum outdoor air supply complies with the IMC. Many HVAC systems combine natural and mechanical ventilation. The natural ventilation strategy offers significant opportunities for energy savings. Mechanical ventilation controls are programmed to maximize the available natural ventilation. This is an efficient approach to space conditioning and indoor air quality.

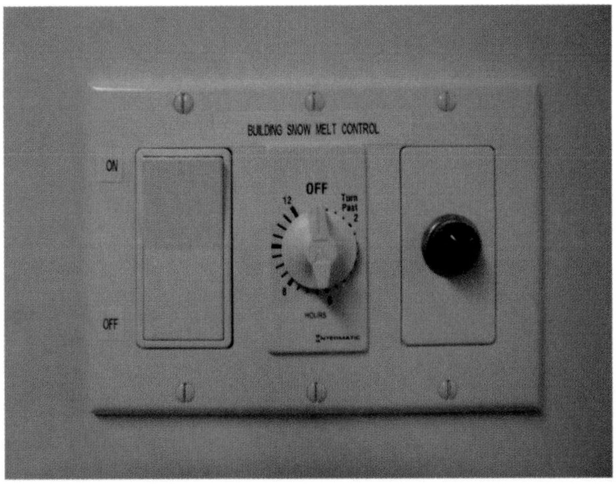

FIGURE 8-3 Manual control is allowed for complete shutdown

Demand-Controlled Ventilation (DCV)

Demand-controlled ventilation (DCV) varies the minimum outdoor air supply to larger assembly areas in the building. DCV systems bring in fresh outdoor air when sensors in the monitored building area measure increased levels of carbon dioxide. Carbon dioxide will accumulate in a room when it is occupied, and the ventilation system may not replace the stale air at a rate to prevent the buildup. The DCV adjusts the air supply based upon the number of people within the space. If used for a small function the system will not fully open and will provide a reduced level of outdoor air while a large function would result in the system opening up and operating near or at its designed capacity. Every zone more than 500 square feet in area with an average occupant load of 25 people per 1000 square feet or more must have a DCV system installed when the HVAC system has an air-side economizer, automatic modulating control of the outdoor air damper, or a design outdoor air flow greater than 3000 cfm. There are exceptions to this provision allowing for specific system components, designs, and controls. **[Ref. C403.2.5.1]**

Energy Recovery Ventilation (ERV)

Energy recovery ventilation (ERV) is an exchange process that transfers the warm or cool temperature in exhaust air and exchanges the energy to temper the supply air coming into the building. Table C403.2.6 of the IECC lists climate zones and percentage of supply air flow rates triggering the ERV requirement. Nine specific exceptions qualify conditions where ERV systems are not required, and include heating energy recovery in climate zones 1 and 2; cooling energy recovery in climate zones 3C, 4C, 5B, 5C, 6B, 7, and 8; for systems serving spaces designed to stay cool and heated to less than 60°F; and for systems operating less than 20 hours a week. **[Ref. C403.2.6]**

ERVs differ from HRVs because they transfer a controlled amount of moisture along with the heat (Figure 8-4). This helps control the humidity in the building. In the summer, an ERV will remove some of the more humid supply air to the exhaust air. In the winter, the ERV will add some moisture to the supply air by removing it from the exhaust air. This transfer is accomplished by an enthalpy wheel.

You Should Know

According to the IMC, the purpose of building ventilation is to protect the health of building occupants by controlling indoor air quality. Outdoor pollution may be introduced to the indoor environment, and indoor contaminants may be produced by equipment, furnishings, and processes. Ventilation is necessary to control these typical air contaminants. Properly designed and functioning ventilation systems are necessary and help create a healthy and comfortable environment for building occupants. ●

FIGURE 8-4 ERVs transfer moisture along with heat

Enthalpy in this context is the exchange of heat *and* moisture. It takes a certain amount of heat energy to change water to steam, just as it takes a certain amount of energy to add water vapor (humidity) to or remove it from the interior air. This moisture transfer, unique to an ERV, is the defining quality of an enthalpy wheel. The U.S. Department of Energy estimates that ERVs are 70–80 percent efficient in recovering energy in the exhaust air and adding it to the supply air.

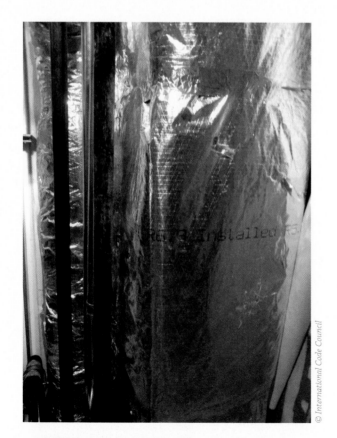

FIGURE 8-5 R-6 is the minimum requirement for ducts and plenums in unconditioned space

DUCTS AND PLENUMS

All air ducts and plenums must be insulated to minimize heat loss and sealed to minimize air leakage. R-6 is the minimum requirement for ducts and plenums in unconditioned space (Figure 8-5). R-8 is the minimum requirement when the duct or plenum is outside the building. The energy code provisions specific to sealing ducts, joints, and seams in ductwork and HVAC components reference IMC Section 603.9. Compliant methods include listed and labeled tapes and mastics and continuously welded and locking joints and seams (Figure 8-6). Ducts, duct fittings, dampers, plenums, and fans must be sealed as required in the energy and mechanical codes. Specific construction requirements for low-pressure duct systems (operating at less than or equal to 2 inches of water gauge pressure), medium-pressure duct systems (more than 2 inches but less than 3 inches of water gauge pressure), and high-pressure duct systems (more than 3 inches of water gauge pressure) are detailed in this section. **[Ref. C403.2.7]**

FIGURE 8-6 Continuously welded and locked joints and seams

Pipe Insulation

Pipes carrying fluids serving HVAC systems must be insulated according to Table C403.2.8 of the IECC. The hotter the fluid moving through the pipe, the more insulation is required. This is partially due to the fact that the cooling loads in commercial building are usually greater than the heating loads. The heat coming off the pipes carrying very hot fluids unnecessarily adds to the heating load if the pipes are not properly insulated (Table 8-2). Note that the IECC table does not require insulation for pipes moving fluids that operate between 60°F and 105°F. This is restated in one of the pipe insulation exceptions. **[Ref. C403.2.8]**

Pipe insulation must be protected when exposed to sunlight, moisture, wind, and possible damage during HVAC equipment maintenance (Figure 8-7). **[Ref. C403.2.8.1]**

TABLE 8-2 Table C403.2.8 Minimum Pipe Insulation Thickness (thickness in inches)[a]

Fluid Operating Temperature Range and Usage (°F)	Nominal Pipe or Tube Size (Inches)				
	< 1	1 to < 1½	1½ to < 4	4 to < 8	≤ 8
< 350	4.5	5.0	5.0	5.0	5.0
251–350	3.0	4.0	4.5	4.5	4.5
201–250	2.5	2.5	2.5	3.0	3.0
141–200	1.5	1.5	2.0	2.0	2.0
105–140	1.0	1.0	1.5	1.5	1.5
40–60	0.5	0.5	1.0	1.0	1.0
< 40	0.5	1.0	1.0	1.0	1.5

[a]For piping smaller than 1 ½ inches (38 mm) and located in partitions within *conditioned spaces*, reduction of these thicknesses by 1 inch (25 mm) shall be permitted, but not to a thickness less than 1 inch (25 mm).

FIGURE 8-7 Pipe insulation must be protected

FIGURE 8-8 Occupancy sensor required for outdoor radiant heating

Commissioning and Completion

Buildings with a total HVAC cooling equipment capacity equal to or greater than 480,000 Btu/h (commonly expressed as 40 tons of cooling) or heating equipment capacity equal to or greater than 600,000 Btu/h must meet the specific requirements in Section C408.2 of the IECC. **[Ref. C403.2.9]** The provisions include a commissioning plan, a final commissioning report, adjusting and balancing the system, testing, and documentation.

Outdoor Heating

Radiant heat systems are required when outdoor spaces are heated (Figure 8-8). An occupancy sensor or timer switch is required so that the heating system turns off when no people are in the space. **[Ref. C403.2.11]**

SIMPLE HVAC SYSTEMS

A simple HVAC system is one designed so that only unitary or packaged equipment supplies the heating and cooling needs for only one zone, and that one zone is controlled by a single thermostat (Figure 8-9). The provisions for these simple systems are limited to economizers in certain climate zones (Table 8-3).

Economizers provide cool outside air to indoor space without the energy cost of producing conditioned air. This is essentially "free cooling" managed by appropriate fan and damper controls and temperature sensors. Economizers are not appropriate in all climates or cost effective for all systems, and specific exceptions for efficiency, hours of weekly operation, and design parameters detail when economizers are not required. **[Ref. C403.3.1]**

Required economizers must be sized to provide all of the indoor cool air from the outside and must include a relief air outlet to prevent building over-pressurization. **[Ref. C403.3.1.1]**

COMPLEX HVAC SYSTEMS

Complex HVAC systems are, well, complex. Any HVAC system that is not unitary or packaged equipment supplying the heating and cooling needs for one zone and that is not controlled by a single thermostat is a

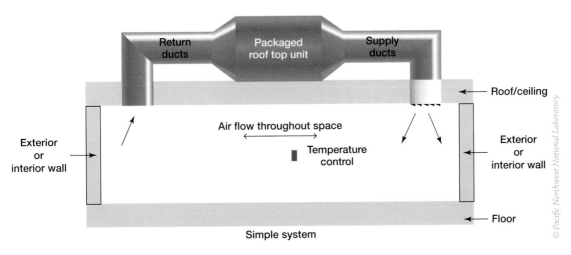

FIGURE 8-9 Simple HVAC system

TABLE 8-3 Table C403.3.1(1) Economizer Requirements

Climate Zones	Economizer Requirement
1A,1B	No requirement
2A, 2B, 3A, 3B, 3C, 4A, 4B, 4C, 5A, 5B, 5C, 6A, 6B, 7,8	Economizers on all cooling systems ≥ 33,000 Btu/h[a]

For SI:1 British thermal unit per hour = 0.2931 W.

[a]The total capacity of all systems without economizers shall not exceed 300,000 Btu/h per building, or 20 percent of its air economizer capacity, whichever is greater.

complex system—basically, any HVAC system that is not simple is complex. Economizers in complex HVAC systems include "cooling towers" common to many large commercial buildings. Chilled water is made during cooler times of the day and stored to circulate through coils during the warmer times of the day to provide "free cooling." **[Ref. C403.4.1]**

Provisions for efficient operation of complex HVAC systems serving multiple zones require controlled variable air volume (VAV) fan systems capable of delivering tempered air only to the zone needing it. This single-zone ventilation design in a multi-zone HVAC system offers significant opportunities for energy savings in any complex building. **[Ref. C403.4.5]**

VAV FANS

Fan motors are an integral component of the HVAC System. They move large volumes of air through ducts and plenums in the building ventilation system. HVAC fans with motors rated greater than 7.5 horsepower must have a variable-speed drive or variable-pitch fan blade controls to reduce the fan motor demand to 30 percent or less at one-half of design air flow. Compliance for this provision is based on the manufacturer's certified data. **[Ref. C403.4.2]**

You Should Know

Electrical power is generally rated in watts or horsepower. A horsepower is a unit of power equal to 746 watts. Most household hairdryers use about 1400 watts to blow warm air to dry your hair. Any VAV fan that uses about as much power as four household hairdryers is required to meet the efficiency requirements of Section C403.4.2. ●

You Should Know

Economizers save the most energy when they are enthalpy controlled. That is the control also accounts for the outdoor air humidity in addition to temperature. ●

Efficient Water Heating

© International Code Council

Commercial building occupants have many uses for hot water. *Service water heating* as defined in the *International Energy Conservation Code* (IECC) is the "supply of hot water for purposes other than comfort heating" **[Ref. C202]** and includes laundries, kitchens, swimming pools, showers, and restrooms. The provisions of this section apply to commercial car washes, recreation centers, and hot water produced in a building system used for any purpose other than comfort heating—the heating in HVAC.

A building's use is a very good indicator of service water heating demand. Table 9-1 is published by the National Renewable Energy Laboratory (NREL) and illustrates estimated hot water consumption by building type.

TABLE 9-1 Water Use Table

Building Type	Estimated Hot Water Use Gallon/Person/Day
House	15.8
Hotel/Motel	20.0
Hospital	52.0
Office	1.1
Restaurant	2.4
School	0.5
School with Showers	1.9

Courtesy: U.S. Department of Energy

EFFICIENCY

Table C404.2 of the IECC (see Table 9-2) lists minimum required efficiencies for water heating equipment. Note that smaller-sized equipment in the same equipment type classification is tested to the DOE 10 CFR Part 430 standard. For more information on the standard, see the DOE section in Chapter 5 of the IECC. This is a reference to the National Appliance Energy Conservation Act (NAECA), an act of Congress that requires this federal standard to be satisfied for all water heating equipment before it can be sold in the United States. Equipment efficiency must be indicated in the mechanical equipment schedule and comply with the performance minimums established in this table. [Ref. C404.2]

You Should Know

The National Appliance Energy Conservation Act (NAECA) of 1987 got its start 12 years earlier as the Energy Policy and Conservation Act (EPCA) in 1975. The EPCA established energy efficiency standards for major household appliances and required the U.S. Department of Energy (DOE) to begin the process of setting efficiency targets. This established national standards for many household appliances. The NAECA continues to evolve, with updates in the Energy Policy Acts of 1992 and 2005 and the Energy Independence and Security Act of 2007.

The NAECA now establishes minimum energy efficiency performance for residential and commercial appliances and consumer products, including office equipment and lighting products. The IECC references the minimum efficiencies for equipment regulated by the NAECA because it is a federal law; allowing anything less is a crime. However, exceeding the NAECA efficiencies is the basis for option 1 in Section C406.1, which lists the additional efficiency package options. ●

TABLE 9-2 Minimum Performance of Water-Heating Equipment

Equipment Type	Size Category (Input)	Subcategory or Rating Condition	Performance Required[a,b]	Test Procedure
Water heaters, electric	≤ 12 kW	Resistance	$0.97 - 0.00132\ V$, EF	DOE 10 CFR Part 430
	> 12 kW	Resistance	$1.73\ V + 155$ SL, Btu/h	ANSI Z21.10.3
	≤ 24 amps and ≤ 250 volts	Heat pump	$0.93 - 0.00132\ V$, EF	DOE 10 CFR Part 430
Storage water heaters, gas	≤ 75,000 Btu/h	≥ 20 gal	$0.67 - 0.0019V$, EF	DOE 10 CFR Part 430
	> 75,000 Btu/h and ≤ 155,000 Btu/h	< 4,000 Btu/h/gal	80% E_t $(Q/800 + 110\ \sqrt{V})$ SL, Btu/h	ANSI Z21.10.3
	> 155,000 Btu/h	< 4,000 Btu/h/gal	80% E_t $(Q/800 + 110\ \sqrt{V})$ SL, Btu/h	
Instantaneous water heaters, gas	50,000 Btu/h and < 200,000 Btu/h[c]	≥ 4,000 (Btu/h)/gal and < 2 gal	$0.62 - 0.0019\ V$, EF	DOE 10 CFR Part 430
	≥ 200,000 Btu/h	≥ 4,000 (Btu/h)/gal and > 10 gal	80% E_t	ANSI Z21.10.3
	≥ 200,000 Btu/h	≥ 4,000 Btu/h/gal and ≥ 10 gal	80% E_t $(Q/800 + 110\ \sqrt{V})$ SL, Btu/h	
Storage water heaters, oil	≤ 105,000 Btu/h	≥ 20 gal	$0.59 - 0.0019\ V$, EF	DOE 10 CFR Part 430
	≥ 105,000 Btu/h	< 4,000 Btu/h/gal	78% E_t $(Q/800 + 110.\ \sqrt{V})$ SL, Btu/h	ANSI Z21.10.3

For SI: °C = [(°F) - 32J/1.8, 1 British thermal unit per hour = 0.2931 W, 1 gallon = 3.785 L, 1 British thermal unit per hour per gallon = 0.078 W7L.

[a]Energy factor (EF) and thermal efficiency (E_t) are minimum requirements. In the EF equation, V is the rated volume in gallons.

[b]Standby loss (SL) is the maximum Btu/h based on a nominal 70°F temperature difference between stored water and ambient requirements. In the SL equation, Q is the nameplate input rate in Btu/h. In the SL equation for electric water heaters, V is the rated volume in gallons. In the SL equation for oil and gas water heaters and boilers, V is the rated volume in gallons.

[c]Instantaneous water heaters with input rates below 200,000 Btu/h must comply with these requirements if the water heater is designed to heat water to temperatures 180°F or higher.

TEMPERATURE CONTROLS

Hot water in the storage vessels and pipe distribution system radiate heat while waiting for the call for use. The energy code provisions for temperature control minimize the potential loss by requiring the water to be heated to no more than the minimum temperature needed to do the job. Hot water is limited to 110°F in public restrooms and dwelling units, and to 90°F everywhere else. The provision only requires the control mechanism to allow these set points and falls short of requiring the water heater to be limited to producing service hot water that exceeds the set points. **[Ref. C404.3]**

The 2012 *International Plumbing Code* (IPC) addresses maximum water consumption for certain fixtures in Table 604.4 (Table 9-3). Conserving water also saves energy. The IPC defines *tempered water* as "water having a temperature range between 85°F and 110°F." *Hot water*

TABLE 9-3 Maximum Flow Rates and Consumption for Plumbing Fixtures and Fixture Fittings-IPC Table 604.4

Plumbing Fixture or Fixture Fitting	Maximum Flow Rate or Quantity[b]
Lavatory, private	2.2 gpm at 60 psi
Lavatory, public (metering)	0.25 gallon per metering cycle
Lavatory, public (other than metering)	0.5 gpm at 60 psi
Shower head[a]	2.5 gpm at 80 psi
Sink faucet	2.2 gpm at 60 psi
Urinal	1.0 gallon per flushing cycle
Water closet	1.6 gallons per flushing cycle

For SI: 1 gallon = 3.785 L, 1 gallon per minute = 3.785 L/m, 1 pound per square inch = 6.895 kPa.
[a]A handheld shower spray is a shower head.
[b]Consumption tolerances shall be determined from referenced standards.

is defined as "water at a temperature greater than or equal to 110°F." IPC Section 416.5 requires *tempered water* to be delivered from fixtures in handwashing sinks in public facilities. The IPC requires an approved temperature-limiting device for sink fixtures in public facilities. IPC Section 424.3 recognizes the need for a good hot shower and allows a water temperature control set point up to 120°F. Section 424.5 allows for a hot bath with the same 120°F temperature control set point.

This is not an example of conflicting provisions in the IPC and IECC but a hint as to how to understand the requirements. The IECC temperature controls address "capable of" set points, whereas the IPC sets "maximum of" allowed temperatures.

HEAT TRAPS

Hot water tends to rise in a storage vessel and can initiate a thermosyphoning loop as cool water replaces the moving warm water. Most commercial water heating equipment have factory-installed heat traps. It is a good idea to specify this if you have the chance. Heat traps are not required on circulating service water heating systems. [Ref. C404.4] Circulating hot water systems require a pump and loop back from the most distant fixture to the water heater. When the faucet is opened hot water is immediately available. The pump can be activated by the push of a button, a timer, or a motion-sensor switch. [Ref. C404.6]

PIPE INSULATION

Water heating piping in circulating systems must be insulated because the pipe is full of hot water most of the time. One inch of

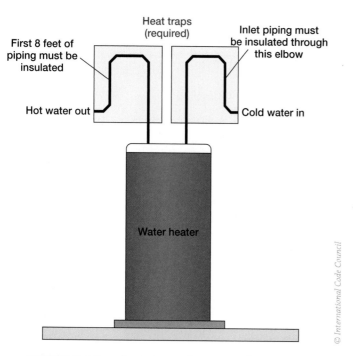

First 8 feet of piping must be insulated

Heat traps (required)

Inlet piping must be insulated through this elbow

Hot water out

Cold water in

Water heater

© International Code Council

FIGURE 9-1 Insulation requirements for water heaters

R-3.5 per inch insulation is required. Piping serving water heaters without an integral heat trap must have ½ inch of insulation installed on the first 8 feet of pipe leading in and out of the water heater (Figure 9-1).

Heat-trace pipe systems have an electric heat tape installed in contact with the water pipe. The heat trace maintains the desired water temperature in the pipe and is an alternate to circulating systems. Insulation requirements are specified by the manufacturer, and any untraced pipe in the system must be wrapped with 1 inch of R-3.5 per inch of insulation. **[Ref. C404.5]**

Both circulating and heat-trace hot water piping systems are required to have a control that will turn off the system when hot water demand is low. **[Ref. C404.6]**

POOL AND SPA HEATERS AND COVERS

Swimming pool and spa operations use energy to heat and filter water. Heat energy is lost when the air temperature at the water surface is lower than the pool or spa water temperature (Figure 9-2). Specific energy-saving provisions apply to pool and spa heaters, time switches, and covers.

Heaters must have an on/off switch mounted outside of the heater that can be reached easily. This allows the heater to be turned off when the pool may not be used for days at a time. Time switches that can automatically

© International Code Council

FIGURE 9-2 Spa cover

turn the heater and pumps on and off according to a preset schedule are required for all installations.

Water evaporates. Pool and spa owners should expect ¼ to ⅜ inch of water loss daily from evaporation, and this is sometimes mistaken for a leak. Evaporation can be significantly reduced with the vapor-retardant pool cover required by the IECC (Figure 9-3). A cover is not required for pools and spas getting more than 70 percent of seasonal heating energy from a heat pump or thermal solar system. **[Ref. C404.7]**

Retractable
pool cover

© International Code Council

FIGURE 9-3 Pool cover

You Should Know

The 2012 IECC provisions only apply to pools and permanently installed in-ground spas. A standard is published that addresses portable electric spas. Jurisdictions adopting the 2012 *International Swimming Pool and Spa Code* find reference to the APSP 14-2011 *American National Standard for Portable Electric Spa Energy Efficiency* in Section 303, titled "Energy." The energy provisions in the pool and spa code are very similar to those in the energy code, with the exception of portable residential spas. Those units must comply with APSP-14 in jurisdictions administering the 2012 *International Swimming Pool and Spa Code*.

The requirements for pool and spa covers are worded differently in the energy and pool and spa codes but accomplish the same intent, which is to prevent evaporation. ●

SECTION
10

Electric Power and Light

© International Code Council

Experts estimate that about 40 percent of a commercial building's energy use is consumed by lighting systems. With this number in mind, the provisions in this section of the IECC are mandatory. Lighting controls offer opportunities for saving energy without the need to change occupant behavior, so in a way the reductions are automatic. Efficient lighting sources and controls maintain and may even improve interior and exterior illumination. Numerous studies suggest that daylighting improves employees' satisfaction with working conditions and reduces absenteeism. Careful attention to lighting design and control benefits building occupants and building owners.

APPLICABILITY

The provisions in this code section apply to lighting systems in new buildings, additions, tenant finishes, alterations to existing lighting systems, and a change in occupancy that increases energy use. Controls and minimum lamp efficiencies are required for interior and exterior lighting systems. Exceptions to the provisions include historic buildings, replacing less than 50 percent of the existing interior luminaries, and changes that do not increase the required lighting power. Lighting in dwelling units within a commercial building is also exempt provided 75 percent of the installed fixtures and lamps are high-efficacy devices. [Ref. C405.1]

LIGHTING CONTROLS

Interior lighting systems must be designed with the intent that lights in the space are on when they need to be and only as many lights are on in the space as needed to make it safe and functional. In areas of a building under skylights and spaces near walls with lots of windows, lighting may not be needed during much of the day. In fact, in these conditions it may be difficult to know if the lights are on at all. These daylight zones must have controls capable of automatically dimming and brightening the space. Exterior lighting control provisions allow for times when the building is used after dusk and before dawn, again to make the areas safe and functional. [Ref. C405.2]

Interior

Interior controls are required in every space enclosed by floor-to-ceiling partitions. The control may be as simple as a user-friendly on/off light switch in the space. An occupancy sensor to activate and deactivate the lights in the room also complies. A control in a remote location is allowable provided a panel shows whether the lights are on or off in the space.

There are two important and logical exceptions to the requirements concerning individual space and occupant-accessible controls: Emergency and security areas that must be continuously lighted and stairways and corridors that are part of the means of egress do not require lighting controls.

Reduction Controls

Spaces required to have manual controls must also have a strategy to reduce the lighting load by 50 percent. The idea is not to have one half of the room bright and the other half dark but to evenly reduce the light in the entire room. The lighting design and fixture schedule will confirm compliance with this provision. Figure 10-2 illustrates three common methods of compliance. One way

You Should Know

High-efficacy lamps are defined as compact fluorescent lamps, T-8 or smaller-diameter linear fluorescent lamps, or lamps with a minimum efficiency of:
1. 60 lumens per watt for lamps over 40 watts;
2. 50 lumens per watt for lamps over 15 watts to 40 watts; and
3. 40 lumens per watt for lamps 15 watts or less.

Figure 10-1 offers efficacy comparisons using this definition and Edison's lightbulb. ●

LAMP	EFFICACY LUMENS/WATT
Edison's first lamp	1.4
Incandescent	10–40
Halogen	20–45
Fluorescent	35–100
Mercury	50–60
Metal halide	80–115
High pressure sodium	100–140

(Courtesy of U.S. Department of Energy, Office of Building Technology, State and Community Programs, www.energycodes.gov)

FIGURE 10-1 Efficacy comparison

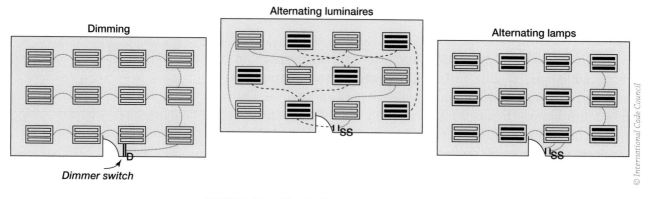

FIGURE 10-2 Typical open office space

is to dim all the lights. This does not work with all types of luminaries. Another approach is to wire two circuits, with each switch controlling half of the light fixtures. The final two-switch option allows one switch to turn on half of the lamps in each fixture and the second switch to turn on the other half of the lamps in the fixtures.

The code notes specific areas and designs that are exempt from the lighting control provisions. These exceptions are areas that have only one luminaire with less than a 100-watt power rating, areas controlled by an occupant-sensing device, corridors and building equipment and service rooms, sleeping units, spaces that use less than 0.6 watts per square foot, and compliant daylight spaces. [Ref. C405.2.1]

Additional Controls

Every area in a commercial building project regulated by the provisions of this code is also required to have one or more of three types of additional controls. Exempted areas are sleeping units, rooms lighted for patient care, areas that could be dangerous when the lights are off, and rooms lighted for continuous use. Automatic time-switch controls satisfy this requirement because they turn off the lighting system when the building space is not in use. The control must be installed so that it is *readily accessible* (see the definition [Ref. C202]) and the lights it controls must be viewable from the switch. Malls, arcades, auditoriums, arenas, some very large buildings, emergency egress lighting, and spaces with occupancy sensors are exempt from these specific provisions.

Occupancy sensors are simple and effectively reduce energy use. These controls turn off the lights after a set time if occupants forget to do so as they leave the room. Sensors are logically required in classrooms, conference and meeting rooms, office break rooms, private offices, storage rooms, janitorial closets, and spaces less than 300 square feet in area with floor-to-ceiling walls. These controls must turn the lights off within 30 minutes of the people leaving. [Ref. C405.2.2.2]

Daylighting design strategy makes the best use of available sunlight by thoughtful and calculated window and skylight placement. These areas are called daylight zones and, when properly designed, require

Code Basics

Readily Accessible: Capable of being reached quickly for operation, renewal, or inspection without requiring those to whom ready access is requisite to climb over or remove obstacles or to resort to portable ladders or access equipment. ●

minimum artificial light. Lighting in daylight zones must be controlled independently of the general lighting system. This allows functional light on bright, sunny days as well as on dark and cloudy ones. The area of a daylight zone depends on the size and number of skylights and windows in or next to the space. A daylight zone can be on any level of the building, not just the areas on the top floor beneath a skylight.

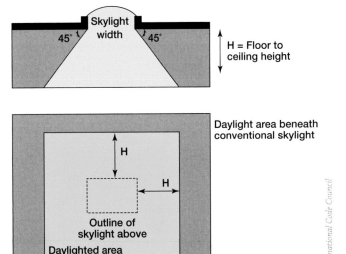

The controls in the daylight zone must be *readily accessible* and have a light-sensing device capable of continuously dimming or dimming in steps the artificial lighting in the daylight zone. If a daylight zone is under a skylight (Figure 10-3) and next to a window, the area must be controlled separately. If a daylight zone beneath a skylight is adjacent to a window's daylight zone, then a single control can be used provided the zone is not facing more than two cardinal orientations. This makes sense because the sun moves across the sky and the interior areas benefit depending on the building's cardinal orientation. If the skylights are located in the interior of the building more than 15 feet from the windows in the perimeter walls, then separate controls for the skylight and window daylighting zones are required. Exceptions are given to areas with floor-to-ceiling walls within a daylight zone and spaces with not more than two light fixtures. **[Ref. C405.2.2.3]**

FIGURE 10-3 Daylight zone under a skylight

Additional controls are required in other specific described instances. Display and accent lighting, such as that used for pictures in hotel or office lobbies and lighting in display cases, must be controlled separately from the other lighting in the space. Hotel and motel rooms and suites must have a control device at the main entry door that operates the permanent light fixtures and switched receptacles in the room or suite. Supplemental task lighting over desks and work surfaces and under counters must have a switch on the fixture or mounted nearby and accessible on the wall. Artificial plant-growth lights and food-warming lights are considered lighting for nonvisual applications. These types of lighting must have a control separate from those of the other lighting in the same room. Lighting fixtures that are displayed for sale and lighting equipment used for education and demonstration also must be controlled separately from the other lighting in the space. **[Ref. C405.2.3]** Exterior building lighting must be connected to a photocell or 24-hour time switch. **[Ref. C405.2.4]**

Interior Lighting Power

The energy code offers two paths of compliance with the interior lighting requirements. Not all lighting equipment is included in the calculations, and most of the exceptions are similar to those in the controls section. Professional indoor sports arenas and fields and casino gaming areas are introduced here as exceptions to the indoor lighting provisions, which makes perfect sense given their size.

You Should Know

Lighting power density (LPD) is not a defined term in the IECC but is a simple concept best understood by its mathematical expression:

LDP = watts divided by the square foot area of the space considered, or w/ft².

LPD is the maximum lighting expressed in the watt rating of each luminaire or multiple lamps used in a single light fixture. This information comes directly from the lighting schedule. The square foot number is the size of the building using the whole-building-area method or the sum of the square feet for the specific space most closely described by the type of activity and LPD number using the space-by-space method. ●

The building-area method limits lighting power density (LPD) by first describing the building and associating the allowable watts per square foot shown in Table C405.5.2(1) in the IECC. This LPD times the building floor area is the interior lighting power allowance. Exempted building area and lighting equipment in those areas are not included in the calculations. The rated wattage of each luminaire is the value used to calculate the proposed lighting power. This information is best described in the lighting plan and lighting fixture schedule.

Example:

The lighting plan for a 1200-square-foot office with no floor-to-ceiling partitions proposes 20 recessed-can light luminaries rated at 50 watts each. The lighting designer specifies 30-watt lamps in each luminaire. What is the LPD for the office space?

Answer: The rated wattage 50 × 20 luminaries = 1000 watts divided by 1200 square feet = 0.83 LPD.

Does this space comply with the space-by-space method?

Answer: Yes. Table C405.5.2(2) allows 1.0 LPD for an open office, and 0.83 is less than 1.0, so the office space complies.

The space-by-space method refers to the building as best described by Table C405.5.2(2) in the IECC (see Table 10-1). Each area of use is associated with a value in the table. The area times the LPD value of all the building spaces equals the interior power allowed for compliance. This method allows trade-offs between spaces to use more or less power to comply if the total building design is less than or equal to the total allowed in the whole building.

TABLE 10-1 Table C405.5.2(2) Interior Lighting Power Allowances: Space-by-Space Method

Common Space-by-Space Types	LPD (w/ft²)
Dining area	
Bar/lounge/leisure dining	1.40
Family dining area	1.40
Dressing/fitting room performing arts theater	1.1
Electrical/mechanical	1.10

TABLE 10-1 Table C405.5.2(2) Interior Lighting Power Allowances: Space-by-Space Method (*Continued*)

Common Space-by-Space Types	LPD (w/ft²)
Food preparation	1.20
Laboratory for classrooms	1.3
Laboratory for medical/industrial/research	1.8
Lobby	1.10
Lobby for performing arts theater	3.3
Lobby for motion picture theater	1.0
Locker room	0.80
Lounge recreation	0.8
Office – enclosed	1.1
Office – open plan	1.0
Restroom	1.0
Stairway	0.70
Storage	0.8
Workshop	1.60
Courthouse/police station/penetentiary	
Courtroom	1.90
Confinement cells	1.1
Judge chambers	1.30
Penitentiary audience seating	0.5
Penitentiary classroom	1.3
Penitentiary dining	1.1
Building Specific Space-by-Space Types	
Automotive–service/repair	0.70
Bank/office–banking activity area	1.5
Dormitory living quarters	1.10
Gymnasium/fitness center	
Fitness area	0.9
Gymnasium audience/seating	0.40
Playing area	1.40
Healthcare clinic/hospital	
Corridors/transition	1.00
Exam/treatment	1.70
Emergency	2.70
Public and staff lounge	0.80
Medical supplies	1.40
Nursery	0.9
Nurse station	1.00
Physical therapy	0.90
Patient room	0.70
Pharmacy	1.20
Radiology/imaging	1.3
Operating room	2.20
Recovery	1.2
Lounge/recreation	0.8
Laundry–washing	0.60

(*Continues*)

TABLE 10-1 Table C405.5.2(2) Interior Lighting Power Allowances: Space-by-Space Method (*Continued*)

Common Space-by-Space Types	LPD (w/ft²)
Hotel	
Dining area	1.30
Guest rooms	1.10
Hotel lobby	2.10
Highway lodging dining	1.20
Highway lodging guest rooms	1.10
Library	
Stacks	1.70
Card file and cataloging	1.10
Reading area	1.20
Manufacturing	
Corridors/transition	0.40
Detailed manufacturing	1.3
Equipment room	1.0
Extra high bay (> 50-foot floor-ceiling height)	1.1
High bay (25–50-foot floor-ceiling height)	1.20
Low bay (< 25-foot floor-ceiling height)	1.2
Museum	
General exhibition	1.00
Restoration	1.70
Parking garage–garage areas	0.2
Convention center	
Exhibit space	1.50
Audience/seating area	0.90
Fire stations	
Engine room	0.80
Sleeping quarters	0.30
Post office	
Sorting area	0.9
Religious building	
Fellowship hall	0.60
Audience seating	2.40
Worship pulpit/choir	2.40
Retail	
Dressing/fitting area	0.9
Mall concourse	1.6
Sales area	1.6

© *International Code Council*

The space-by-space method table ends with a very important footnote about the sales area in retail space. The basic allowance of 1.6 watts per square foot for a sales area may be increased by the area allotted to specific classes of merchandise according to the formula in the footnote section. When these allowances are taken, the additional lighting circuits must be switched or dimmed separately from the general lighting. [**Ref. C405.5.2**]

EXTERIOR LIGHTING

Lighting installed in parking lots, outside building entrances, on the walkways connecting the parking areas, and highlighting building façades is regulated by the provisions in this section. Lighted advertising and traffic directional signs are exempt, as are solar lights not connected to the building power source. Lighting for public monuments (Figure 10-4) and exterior features of historic landmark structures and buildings is also exempted.

FIGURE 10-4 Exempt exterior lighting

The basic power allowance and LPD for specific exterior lighting applications is determined by Table C405.6.2(1) in the IECC, which describes zones (Table 10-2). After choosing the most appropriate zone considering the building's location, calculate the total exterior lighting power according to Table C405.6.2(2).

Tradable surfaces are like the whole-building LPD calculation. Once the allowed total is established, the designer may use more light in one area and less in another, and still comply if the total is equal to or less than the budget allowed. Nontradable surfaces generally include security lighting that needs to be on all night, such as at ATMs, ambulance entries, and inspection stations at guarded entries. Building façade lighting is exempt because sometimes people need to find facilities at night. In all exterior lighting applications, luminaries rated to operate at more than 100 watts must have an efficacy greater than 60 lumens per watt. **[Ref. C405.6]**

TABLE 10-2 Table C405.6.2(1) Exterior Lighting Zones

Lighting Zone	Description
1	Developed areas of national parks, state parks, forest land, and rural areas
2	Areas predominantly consisting of residential zoning, neighborhood business districts, light industrial with limited nighttime use and residential mixed use areas
3	All other areas
4	High-activity commercial districts in major metropolitan areas as designated by the local land use planning authority

Required Efficiency Options

© International Code Council

ADDITIONAL EFFICIENCY FEATURES

The additional efficiency provision of Section C406 in the *IECC* completes the prescriptive path of energy code compliance for commercial buildings. Three options complement the minimum mandatory provisions for building envelope performance and mechanical systems, service water heating, and lighting efficiency. **[Ref. C406.1]** This requirement effectively "raises the bar" for the prescriptive compliance path. The ASHRAE 90.1-2010 and the performance path (Section C401.2 Items 1 and 3) building designs produce similar overall results in effective and efficient energy use.

HVAC System Performance

Tables C406.2 (1–7) include higher efficiency requirements for most of the equipment system types listed in Tables C403.2.3 (1–7). If this option is chosen, the provisions for mechanical systems [Ref. C403] still apply and the HVAC equipment efficiency must comply with the applicable equipment choices in Tables C406.2 (1–7). This option shall only be used where the equipment efficiencies in Tables C406.2 (1–7) are greater than the general requirements in Tables C403.2.3 (1–7).

Lighting

This option reduces the lighting power density (LPD) to the values in Table C406.3 versus those in Table 405.5.2(1). Footnote (b) offers daylighting control (Figure 11-1) as an option for compliance in retail and office space. Footnote (c) requires at least 70 percent of warehouse floor area to be in the daylight zone. Table C406.3 does not allow additional lighting in retail spaces. Compliance requires that the building space functions are identified and measured, and the supporting LPD calculations must be submitted with the plans to demonstrate compliance.

Onsite Renewables

Two renewable system choices are available to comply with option 1 of this provision. A photovoltaic (PV) system (Figure 11-2) providing at least 0.50 watts or a solar thermal (hot water) system providing at least 1.75 Btus per square foot of the building conditioned floor area satisfies this option.

Option 2 requires that at least 3 percent of the energy used by the mechanical, service hot water, and lighting systems in the building be provided by onsite renewables. This allows flexible design to maximize the local solar resources to fit the building energy demands.

FIGURE 11-1 Daylighting control

FIGURE 11-2 Photovoltaic (PV) and solar thermal (hot water) systems

GENERAL RESIDENTIAL ENERGY PROVISIONS

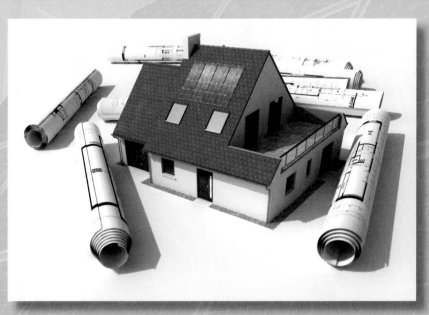

© Franck Boston/www.Shutterstock.com

General Provisions

© International Code Council

The provisions of R101 and R102 of the *International Residential Code* (IRC) establish a framework to address the administration, application, and enforcement of the residential energy code provisions. These sections address requirements for organization of the building plans and information needed in the supporting documents. Many states require the construction documents to be prepared by a registered design professional, usually an architect or engineer. Homeowner builders may be allowed to submit building plans and build their own homes. It is advisable to check with the code official in the jurisdiction to get specific information about this process.

The building contractor's responsibilities regarding permits and inspections and the code official's responsibilities concerning plan review and inspections are individually addressed and detailed in these sections. These sections of the IRC describe the relationships and understandings between the building department authority and the design and construction community.

SCOPE

The residential provisions apply only to residential buildings. This distinction is not as obvious as it seems. Entire buildings and building sites or just parts of buildings systems and building equipment may be classified as residential. [Ref. R101.2] It is essential to know the difference in residential and commercial classification. As stated in R202, the definition of *residential building* is as follows: "For this code, includes detached one- and two-family dwellings and multiple single-family dwellings (townhouses) as well as Group R-2, R-3 and R-4 buildings three stories or less in height above grade plane." *Grade plane* is defined in Section 202 of the *International Building Code* (IBC). As noted earlier in this text, an italicized term in the code means that the word or phrase has a specific meaning in code language, and a definition is provided to clarify its intended use. Chapter 2 of the IECC lists 58 words and specific definitions to establish the common vocabulary for the residential energy regulations.

Classifications and descriptions of residential occupancies are found in IBC Section 310. Apartments, condominiums, and timeshare buildings are typical R-2 occupancies. Single-family homes, duplexes, and townhomes are R-3 buildings. The shared attribute of these residential uses is that the occupants are "non-transient." Thus, if the building uses fit into any of the defined descriptions of "residential" and the building is less than three stories in height, the IECC residential provisions apply. [Ref. R101.2]

INTENT

As noted in Section 3 of this text, the intent of the IECC is stated simply: "This code <u>shall</u> regulate the design of and construction of buildings for the effective use and conservation of energy over the useful life of each building." *Webster's Dictionary* defines *shall* as "used in laws, regulations, or directives to express what is mandatory." All the provisions and regulations in the energy code must pass this test of intent to conserve energy use in the design and construction of buildings to be included in the code. The design requirements of the energy code are intended to promote the effective use of energy in new buildings and building systems as well as in repairs and other work undertaken in existing buildings, "over the useful life of each building." As noted earlier in the discussion of the IECC's commercial provisions, "useful life" is a somewhat ambiguous term, as many factors may affect a building's life span. In the case of residential buildings, many are only meant for a useful life of 30 years or so, the length of a typical mortgage, whereas others have existed for a century or more and may be classified as historical buildings.

Also, as noted in the commercial provisions discussion (Section 3), the intent of the code is to encourage—not stifle—the use of new and innovative energy-saving devices and designs, and therefore materials and techniques not specifically addressed in the IECC may be approved by the code official if they contribute to the overall intent of conserving energy.

APPLICABILITY

As in the commercial provisions, the residential regulations of the IECC note both general and specific requirements for certain conditions. In general, when two different code provisions apply to the same condition, the most restrictive requirement applies. However, in situations where the difference is between a general requirement and a specific requirement, the specific provision applies to the design and construction, as is true of the other I-Codes. **[Ref. R101.4]**

EXISTING BUILDINGS AND HISTORICAL BUILDINGS

The code provisions for existing and historical residential buildings are the same as those for existing and historical commercial buildings and will be briefly summarized here; please see Section 3 of this text for a complete discussion of these regulations. Basically, the IECC does not require retroactive alterations to existing buildings or building systems to bring them into compliance with the minimum energy code provisions, but the IECC provisions do apply to buildings constructed without a permit and most types of work proposed for an existing building that require a permit. **[Ref. R101.4.1]**

Existing buildings may also be classified as currently listed, designated, or certified as historic or in the process of certification, in which case the historic building provisions of the IECC apply. Basically, all aspects of alteration or repair of historic buildings are exempt from the energy code. (See Section 3 for a complete discussion.) Although a historic building is not required to comply with the IECC, the owner could elect to use the guidance of the code provisions in order to make the building more energy efficient; the key distinction is that the IECC is an option rather than an enforceable requirement. **[Ref. R101.4.2]**

PERMITS

The IECC does not define the specific "energy code" permits that are required. However, the IECC does define the work that must comply with the minimum code requirements and the types of work that require a permit to meet compliance. Portions of an existing building undergoing renovation or repair (but not the unaltered or unaffected portions of the building, which are exempt) must comply with energy code provisions required for new construction. Any addition to an existing building also must comply with energy code provisions required for new construction.

Certain types of work, however, do not require compliance, such as proposed construction or alterations that will not increase building energy use. For example, storm windows installed over existing windows improve thermal performance and reduce energy use. This popular homeowner project is specifically exempted from the energy performance requirements for new windows. The code also exempts replacing broken glass in existing window frames. This exception does not, however, exempt whole window replacements. Installing a new window assembly in an existing window opening requires the window to meet the same solar heat gain coefficient (SHGC) and U-factor requirements as would be required for new construction. Typical energy-upgrade projects that include window replacement require permit application, plan review, permit issuance, and inspection. [Ref. R101.4.3]

You Should Know

Owner-Occupied Lodging Houses is a new residential category that has been added in the 2012 IRC. These are commonly known as bed-and-breakfast and can be constructed under the provisions of the IRC if they have 5 or fewer guestrooms and are equipped with a fire sprinkler system in accordance with IRC Section P2904." ●

CHANGE IN OCCUPANCY OR USE

Changes in residential tenant occupancy seldom trigger energy code compliance. Apartment tenants and renters come and go and homes are remodeled, but the classification of "residential occupancy" is unaffected. [Ref. R101.4.4]

For residential buildings, a change in use most often happens when any unconditioned space becomes heated or cooled, and this may trigger code compliance. For example, a garage, outbuilding, storage area, or basement without heating or cooling that becomes a den, bedroom, apartment, or living area must comply with all the provisions of the code. [Ref. R101.4.5]

All residential buildings—including multi-family residential buildings not more than three stories high—are regulated by the provisions in this part of the code. As noted in the commercial provisions discussion, in buildings not more than three stories tall that contain both commercial spaces and residential units, the residential units must comply with these code provisions. Hotel or apartment buildings four stories or more, even though the entire use is residential, must meet the commercial energy provisions. [Ref. R101.4.6]

OTHER RESIDENTIAL EXEMPTIONS

Low-energy residential buildings are exempt from the thermal envelope code requirements. The building or portion of the building must be separated from the conditioned areas with walls or ceilings that comply with the thermal envelope provisions. For example, an attached garage with a space-heating unit intended to keep exposed pipes from freezing may qualify for this exemption. The maximum heating capacity allowed is 1.0 watt/ft^2. Low-energy buildings are not exempt from the mandatory lighting, heating, and service hot water provisions of the code. [**Ref. R101.5.2**]

COMPLIANCE

The compliance path for residential buildings is the same as that for commercial buildings, and will be briefly reviewed here; please see Section 3 of this text for a complete discussion. Basically, whether new construction, alteration, repair, or an addition, the plans, specifications, and details must comply with the minimum provisions of the code. The simplest compliance path for single-family and noncomplex residential structures is to use the prescriptive worksheets offered by some jurisdictions that establish the requirements applicable to that location. Another option, discussed fully in Section 3, is the U.S. Department of Energy's RES*Check*™ software program. Most, but not necessarily all, jurisdictions will accept ResCheck documentation. It is always best to talk with the building official before using any type of software or worksheets and determine if the product and its results are acceptable to the jurisdiction. [**Ref. R101.5.1**]

ALTERNATE MATERIALS

As noted, the intent of the code is to encourage innovative designs, materials, and techniques for energy savings; thus, any material or method of construction that meets the intent of the applicable code provision may be evaluated and approved by the code official. [**Ref. R102.1**] In addition, many jurisdictions adopt, administer, and enforce mandatory above-code programs, which may be deemed compliant by the code official. However, as noted in Section 3, even when an alternate program is deemed to comply, the mandatory energy efficiency provisions in Chapter 4 of the IECC must be met. [**Ref. R102.1.1**]

You Should Know

The ICC Evaluation Service (ICC-ES) in an independent body that evaluates and issues reports of compliance based on various codes and standards. ICC-ES Evaluation Reports (ESRs) are available to all users free of charge and provide a valuable tool for the safe and consistent use of construction materials. ICC-ES also provides evaluation of sustainability attributes based on energy codes, green codes, standards, and various rating systems, and also has building products and plumbing/mechanical/fuel gas (PMG) listing programs in support of the construction industry professionals. ●

Administration and Enforcement

© International Code Council

PREPARING THE PLANS

As with commercial buildings, for residential buildings the permit applicant is responsible for preparing a complete set of construction documents for the building project. In many cases a set of building plans with sufficient details and schedules will describe a simple project and demonstrate compliance with the energy code. A more complete understanding of the code provisions and requirements by the homeowner-builder, designer, and contractor will help the building department plans examiner review the project for compliance and speed the process of permit issuance. It is important to determine if the jurisdiction requires the plans to be prepared by a registered design professional. The local code official will have information about the applicability of this requirement to specific projects, and is also authorized to waive this requirement if it is determined that a registered design professional's stamp and seal is not required to confirm compliance with the IECC. **[Ref. R103.1]**

Elements of the building envelope, mechanical equipment, duct sizing, and hot water pipe and duct insulation information must be included in the plans for energy code compliance review. **[Ref. R103.2]**

The prescriptive or performance compliance path chosen will determine the detail required to demonstrate code compliance. The prescriptive provisions provide a simple path to building document preparation.

THE CONSTRUCTION DRAWINGS AND DOCUMENTS

The Thermal Envelope and Air Sealing

As noted earlier in Section 4, the plans must include building elevations depicting window, door, and skylight areas and wall sections showing the type of insulation and its thickness. A good trick to remember is to use your finger to trace the thermal envelope on the drawing sections and details (Figure 13-1). The design documents should have sufficient detail to trace all six sides of the building cube without lifting your finger

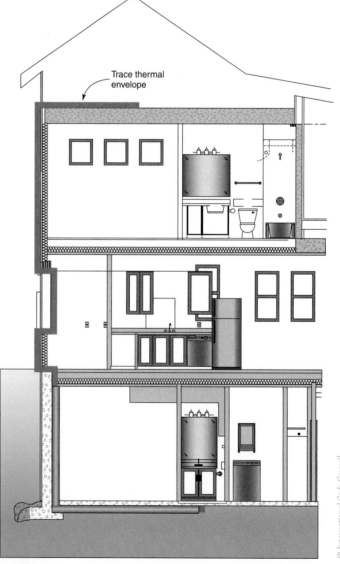

Trace thermal envelope

© *International Code Council*

FIGURE 13-1 The thermal envelope must be completely described by the drawings

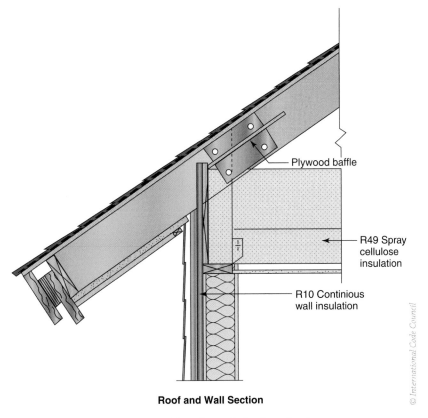

Roof and Wall Section

FIGURE 13-2 R-value of the insulation

off the paper. Three sides of the cube are shown in this figure. This trick also works to confirm the air-sealing continuity in the design drawings.

The wall and roof sections in the plans must list the R-value of the insulation (Figure 13-2), and the window and door schedules must list the U-factors and solar heat gain coefficient (SHGC) of the fenestration specified for the project. Above- and below-grade wall construction and framing must be identified separately.

The insulation and material requirements differ for wood- versus metal-framed walls. Building plan elevations indicating window and door sizes and locations, and a roof plan showing skylights, must agree with the schedules, notes, and callouts. Caulking and sealing details for window and door frame installation must be included in the construction documents submitted for plan review. Figure 13-3 shows notes and details that are typical of how all exterior joints are to be sealed.

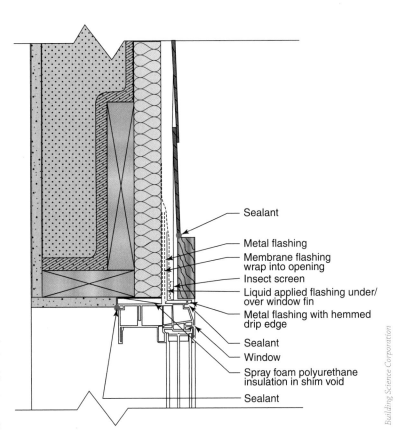

FIGURE 13-3 Details on how exterior joints should be sealed

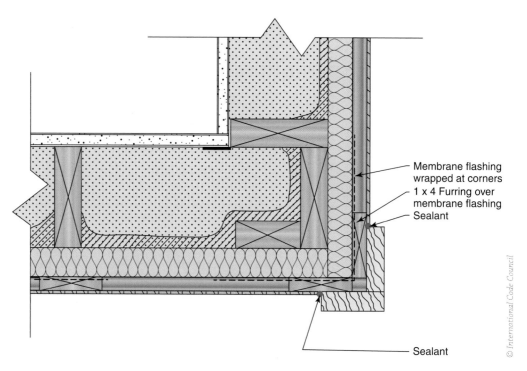

Membrane flashing
wrapped at corners
1 x 4 Furring over
membrane flashing
Sealant

Sealant

FIGURE 13-4 Exterior joint caulk and seal

Air infiltration and exfiltration must be addressed in the building plan documents. All exterior joints, cracks, and holes are to be gasketed, caulked, weather-stripped, or sealed (Figure 13-4). The air-sealing materials must be called out on the plans and compatible with the building materials in the joints and holes where they are applied. Details, notes, and sections in the building plans must address the continuity of the air sealing for the entire building envelope.

Building Systems

Mechanical system design must be included in construction documents. Some residential building HVAC and plumbing systems are sophisticated and complex, and the documents must be prepared by registered design professionals. Mechanical and plumbing systems in single-family homes may be simple compared to those for a larger residential project, and may be prepared by someone other than a registered design professional. The documents, whether submitted by a licensed engineer or a homeowner-builder, require mechanical equipment and duct sizing calculations and HVAC schedules that include the types and efficiencies of the mechanical equipment. Duct-sealing materials must be specified to demonstrate compatibility with the duct materials and the ducts are required to be insulated. Certain service hot water pipes are required to be insulated. High-efficacy lighting fixtures and lamps must be specified on the plans as well. This information is typical of the details, callouts, and drawings required for a complete and compliant set of construction documents.

PLAN REVIEW

Just as with commercial buildings, the code official is charged with reviewing the construction documents—including schedules, calculations, and specific product information—for residential buildings for code compliance. [Ref. R103.3] The final step in permit issuance is for the official to stamp the approved documents "Reviewed for Code Compliance." As with commercial projects, one set of plans is kept in the building department office and an identical set is kept at the work site. This set of construction documents cannot be altered or changed without the code official's approval. [Ref. R103.3.1]

The work site set is the reference for field inspections, and the construction must agree with the approved plans. Any changes to the approved plans must be resubmitted to the code official for further review, approved by the official, and returned to the field before inspections are called for that portion of the work. [Ref. R103.4]

Moisture and vapor control strategy must be addressed in the building plan details. The requirements in Section R702.7 of the 2012 *International Residential Code* (IRC) (Figure 13-5) must be incorporated in the building plan's wall section details. Exterior walls must be covered with materials approved as weather coverings, and in certain climate zones they must include a vapor retarder. It is important to be familiar with and incorporate these provisions in the plans (Figure 13-6).

> **R702.7 Vapor retarders.** Class I or II vapor retarders are required on the interior side of frame walls in climate zones 5, 6, 7, 8 and Marine 4.
>
> **Exceptions:**
> 1. Basement walls.
> 2. Below grade portion of any wall.
> 3. Construction where moisture or its freezing will not damage the materials.
>
> **R702.7.1 Class III vapor retarders.** Class III vapor retarders shall be permitted where any one of the conditions in Table R702.7.1 is met.
>
> **R702.7.2 Material vapor retarder class.** The vapor retarder class shall be based on the manufacturer's certified testing or a tested assembly.
>
> The following shall be deemed to meet the class specified:
>
> Class I: Sheet polyethylene, unperforated aluminum foil.
> Class II: Kraft-faced fibreglass batts.
> Class III: Latex or enamel paint.

FIGURE 13-5 Moisture and vapor control must be addressed in building plan details

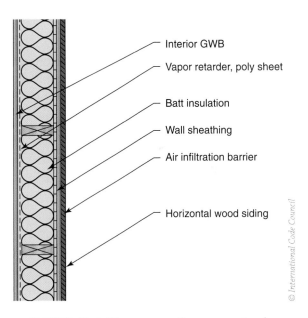

Interior GWB

Vapor retarder, poly sheet

Batt insulation

Wall sheathing

Air infiltration barrier

Horizontal wood siding

FIGURE 13-6 Vapor retarder as required

INSPECTIONS

As with commercial projects, residential building construction work that requires a permit also requires inspection. Specific energy code inspections are not listed in the IECC as they are in the other I-Codes. The periodic inspections required by the normal construction process will incorporate many of the energy code provisions in the inspection schedule. For example, the foundation inspection requires perimeter and below-grade wall insulation to be in place before backfilling, and fenestration compliance and air sealing are verified in the framing inspection. The insulation inspection verifies the required material R-values are in place and properly installed.

Mechanical and plumbing inspections are required to confirm compliance with the approved plans. The mechanical plans and schedules indicate the mechanical equipment efficiency and duct size, location, and

Code Basics

Section 110.3.7 of the *International Building Code* (IBC), "Energy Efficiency Inspections," requires compliance for buildings that are constructed under the provisions of the IBC. The inspections include but are not limited to envelope insulation R-value, fenestration U-value, duct system R-value, and HVAC and water-heating-equipment efficiency. It is the "shall include, but not be limited to" language in this inspection requirement that requires special attention.

Most state and local jurisdictions amend the codes during the adoption process; thus, it is necessary to consult the authority having jurisdiction to find out exactly what the inspection requirements are to confirm compliance with the energy code. The IRC does not include a specific requirement for energy efficiency inspections. •

☑ Building thermal envelope. Specific items shall be inspected and approved.

☑ Fenestration. All fenestration as defined in Section 202 shall be labeled, inspected and approved before the gypsum board inspection.

☑ Insulation. All above and below grade wall and ceiling cavity insulation shall be labeled, inspected and approved before the gypsum board inspection.

☑ Other inspections. In addition to the inspections specified above other inspection shall include, but not be limited to, inspections for: duct system *R*-value, HVAC and water-heating equipment efficiency and lighting compliance.

☑ Final inspection. The building shall have a final energy efficiency inspection and not be occupied until *approved*.

FIGURE 13-7 Typical residential energy code inspections

insulation. Plumbing inspections check that the equipment efficiencies and pipe insulation is installed as required and as approved on the plans. The final building inspection includes counting the number and type of high-efficacy light fixtures and lamps to verify energy code compliance. **[Ref. R104.2]**

Specific energy code inspections are usually a matter of building department policy and procedure. A final inspection is the only required inspection mentioned in the IECC. **[Ref. R104.3]** This implies that all other field verifications for compliance blend with the community standards regarding the normal sequence of inspection in the jurisdiction. Contractors and tradespeople must be responsible for knowing the local rules regarding field inspection expectations (Figure 13-7).

FEES

As covered in Section 4, jurisdictions establish a fee schedule when they adopt code provisions. Fee amounts vary, so specific information must be gathered from the jurisdiction responsible for the proposed project. Any and all applicable fees must be paid before the permit is issued. Energy code compliance may require building, mechanical, plumbing, and electrical permits. Any work that needs a permit should not begin before the applicable permit is issued. **[Ref. R107.1]** Starting work without a permit may trigger additional fees to be added to the permit fee. **[Ref. R107.3]** It is not a good idea to start work before a permit is issued because it may not comply with the applicable code provisions—resulting in more time, money, and materials than necessary.

ENFORCEMENT

A stop work order is the ultimate enforcement tool. The code official has the authority to stop work on any part of a project if the work regulated by the code is dangerous, unsafe, or not code compliant. [Ref. R108.1] The stop work order is issued in writing to the owner or person doing the work and is specific to the noncompliant work that must be stopped. [Ref. R108.2] If the stop work order is ignored, the code official may levy a fine, an action that generally involves a citation in municipal court. The amount of the fine is set by each jurisdiction as determined at the time of code adoption. [Ref. R108.4]

BOARD OF APPEALS

As with commercial construction, the board of appeals exists to allow a means to appeal decisions concerning residential construction made by the building official. Members of appeal boards are usually civic-minded architects, engineers, insulators, lighting designers, or mechanical, plumbing, or electrical contractors. [Ref. R109.3] The board of appeals decides whether the appellant or the code official's action is consistent with the intent of the code. In some cases, the energy code may not apply to a specific project condition; in others, an alternative approach to a proposed construction method may be rejected by the code official because it does not appear in the code, but the board may deem that the method or material is equal to that described in the code.

Note that it is the *intent* of a certain code provision that generates the basis for an appeal. An appeal can be heard when an applicant, designer, or tradesperson does not agree with the code official's decision or order. The appeal may concern a design or field inspection decision and be based on an opinion that the true intent of the code provision or rule has been incorrectly applied by the code official. [Ref. R109.1] The board hears the appeal, makes a decision, and records the findings in writing. In no case may the board waive any requirements of the code. [Ref. R109.2]

Code Basics

Zone: A Space or group of spaces within a building with heating or cooling requirements that are sufficiently similar so that desired conditions can be maintained throughout using a single controlling device. •

DEFINED TERMS

The 58 specific words and terms defined in the IECC have a common and agreed-upon meaning when written and appearing in the code provisions. [Ref. R201] For example, the words *accessible*, *readily accessible*, *building*, *listed*, *repair*, *basement wall*, and *zone* all are defined. Any defined word or term appears in *italics* in the code text. In reading, interpreting, and applying code provisions, it is essential to use the common code vocabulary. [Ref. R202] This is the basis for civil discussion and deeper understanding of the energy code provisions.

PART

V

Specific Requirements for Residential Buildings

© Alena Brozova/www.Shutterstock.com

Specific Requirements for Residential Buildings

© International Code Council

GENERAL

An Energy Efficiency Certificate listing installed insulation, glass and door ratings, and heating and cooling equipment efficiency and test results must be posted on or in the electrical panel in the dwelling. The certificate is useful consumer information comparable to the option package and estimated miles-per-gallon sticker on new car windows. It creates a permanent record of building elements that are difficult to determine after the construction is finished.

The certificate must include building envelope, duct insulation, and window and door U-factor and solar heat gain coefficient (SHGC) values. The window and door information is valuable when replacement

2012 IECC Energy Efficiency Certificate

Insulation rating	R-Value	
Ceiling/Roof	59.00	
Wall	30.00	
Floor/Foundation	15.00	
Ductwork (unconditioned spaces):	_____	

Glass & door rating	U-Factor	SHGC
Window	0.28	0.40
Door	0.35	NA

Heating & cooling equipment	Efficiency
Heating system: _____	_____
Cooling system: _____	_____
Water heater: _____	_____

Building air leakage and duct test results	
Building air leakage test results	_____
Name of air leakage tester	_____
Duct tightness test results	_____
Name of duct tester	_____

Name: _____ Date: _____

Comments:

© International Code Council

FIGURE 14-1 Energy Efficiency Certificate template

projects are contemplated. HVAC system and water heater efficiencies are required to be listed. This information is a guide to maintaining or increasing energy savings when any of the equipment is replaced. The blower door and duct leakage test results are also recorded. The whole house number establishes the baseline with which to compare future test results to evaluate air barrier performance over time (Figure 14-1).

The builder or registered design professional is responsible for completing the information required on the certificate. **[Ref. R401.3]**

Specific Requirements

© International Code Council

ABOUT THE THERMAL ENVELOPE

The building thermal envelope separates the interior from the exterior and encloses conditioned space. Building elements of the thermal envelope include basement and exterior walls, roofs, windows, doors, skylights, and floors. The key to establishing the boundaries of the thermal envelope is found in the definition of *conditioned space*. As defined in the *International Energy Conservation Code* (IECC), *conditioned space* is "an area or room within a building being heated or cooled, containing uninsulated ducts, or with a fixed opening directly into an adjacent conditioned space." It is important that the project designer define the limits of the building thermal envelope in the plans and demonstrate compliance with the mandatory and applicable requirements of Chapter 4 of the IECC. In some designs, crawlspaces, attics, and areas behind knee walls are unconditioned and exposed directly to the outdoors, separated only by uninsulated exterior walls or roofs.

CODE COMPLIANCE PATHS

The code offers three choices for prescriptive compliance (Figure 15-2). Table R402.1.1 offers a very simple and straightforward compliance method. Individual components of a typical building envelope are listed and the performance level of each is established by climate zone. Not every building design will use every component, and it is important to be familiar with the definitions and descriptions of these components. Only then can the required minimum R-values for the ceiling and insulation components for the wood-frame, basement, or crawlspace walls and

You Should Know

The U.S. Census Bureau reports that in 2010, there were 131,704,731 existing housing units in the United States.[1] The census data summarized in Figure 15-1 show that about 60 percent of these homes were built before 1980. The wall insulation values for these existing homes are not separated by climate zones but are well below values in the current energy code provisions[2] (Figure 15-1). The energy code provisions require areas in a building undergoing an alteration, renovation, or repair to conform to the provisions for new construction. The ultimate intent of this code requirement is that over time, with few exceptions, the wall insulation in existing homes and buildings will conform to the current energy code provisions. **[Ref. R101.4.3]** •

[1]U.S. Bureau of the Census. 2012. *Census of Population and Housing, Profiles of general Demographic Characteristics*. http://quickfacts.census.gov.
[2]"Exterior Wall Insulation." Madison Gas & Electric. http://tinyurl.com/MGE-Exterior.

Year of construction	Percent of total housing units*	Wall insulative R-value
Before 1950	20.40%	4
1950–1960	11.50%	3 or 6.2
1960–1970	11.60%	6.2 or 11
1970–1980	16.70%	11 and 5†
1980 to present	39.90%	11 and 5;† 19 and 5†,‡

* There are roughly 127.7 million homes in the U.S., according to data taken from 2005–2009.
† "and" indicates use of multiple types of insulation.
‡ Each set of numbers reresents the R-value for a common insulation combination used.

Table 1: U.S. housing stock and typical insulation R-values per construction year.

Window type	Annual cost
Single-glazed with storm window	$1,310
Double glazed	$1,228
Double glazed with low-e	$1,120
No window (Insulated walls only)	$1,000
Super window	$960

Table 2: Heating costs (average house in a heating climate) with different window types.

© International Code Council

FIGURE 15-1 Fenestration and wall insulation

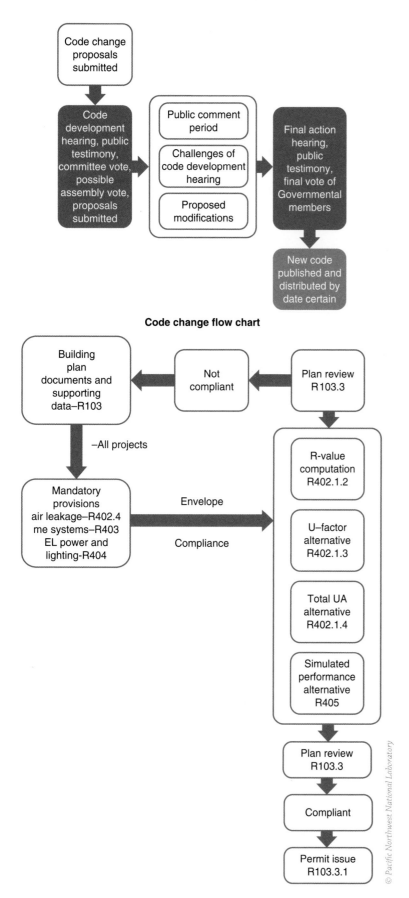

Code change flow chart

FIGURE 15-2 Plan review flow chart

© Pacific Northwest National Laboratory

U-factor and solar heat gain coefficient (SHGC) values for windows and skylights be determined. Lower-level house designs may include combinations of all of these elements. When building components are clearly identified in the drawings, the minimum R-values can be specified for the required insulation and the U-factor and SHGC values for the windows and skylights.

Only the R-value of the insulation products (Table 15-1) is used to confirm compliance with the requirements of this table; no consideration is given for any component other than the insulation products to determine the R-value of the opaque assemblies in the thermal envelope component.

The footnotes to this table offer design choices and specific direction regarding insulation placement. Footnote h allows wood-frame walls in climate zones 3 through 5 to be constructed with R-20 cavity insulation or R-13 cavity insulation plus R-5 continuous insulation. This allows the designer or builder to choose 2 × 6 or 2 × 4 stud wall construction. **[Ref. Table R402.1.1]**

Table R402.1.3 is the U-factor version of Table R402.1.1 and offers the U-factor computation method in place of the R-values. All parts of the construction assembly (Figure 15-3) that contribute to lessening thermal transfer are given a value and considered in the U-factor computation (Figure 15-4). This method is advantageous when using construction methods that limit framing (24" OC versus 16" OC) or include thermal breaks. Resource material outside the IECC must be consulted to determine compliance with this method. The ASHRAE *Handbook of Fundamentals* is often used to calculate the U-factor.

Section R402.1.4 offers a tradeoff path to compliance. This UA alternative allows a design for building components that do not comply with the

TABLE 15-1 Insulation R-Value Primer Table

Fiberglass Batt Insulation Characteristics	
Thickness (inches)	**Typical R-Value**
3 ½	11
3 ⅝	13
3 ½ (high density)	15
6 to 6 ¼	19
5 ¼ (high density)	21
8 to 8 ½	25
8 (high density)	30
9 ½ (standard)	30
12	38
Foam Plastic Insulation Characteristics	
Polystyrene	R-3.8 to R-5 per inch
Polyisocyanurate	R-5.6 to R-8 per inch
Polyurethane	R-7 to R-8 per inch
Cellulose is a loose-fill insulation material provides a thermal resistance of R-3.6 to R-3.8 per inch.	

© International Code Council

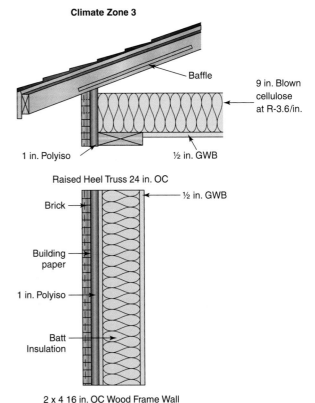

Climate Zone 3

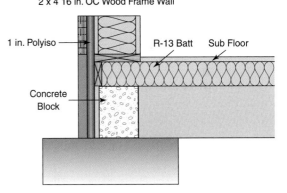

FIGURE 15-3 U-value

© International Code Council

requirements of Table R402.1.3. Calculations must demonstrate that the total UA is the same as or less than that of a building of the same design that complies with the equivalent U-factor method in Table R402.1.3. Supporting calculations simply multiply the area of the component (walls, roof, windows, and doors) times the proposed design U-factor. The same component area is multiplied by the U-factor required in Table R402.1.3. The sum of the proposed design must be less than the standard code design for compliance.

The *simulated performance alternative* is the third path to energy code compliance. This method uses software to predict the annual energy use of the proposed design compared to that of the standard design of building components listed in Table R405.5.2(1). If the proposed building's annual energy used is less than or equal to that of the standard design, the building complies with the code's goal of effective energy use. This method is useful for advanced building designs that incorporate shading strategies to minimize air-conditioning loads in the summer and maximize heat gain in the winter. Window orientation, efficient mechanical systems, and on-site renewable energy sources are allowed input values in this sophisticated compliance method. **[Ref. R405]**

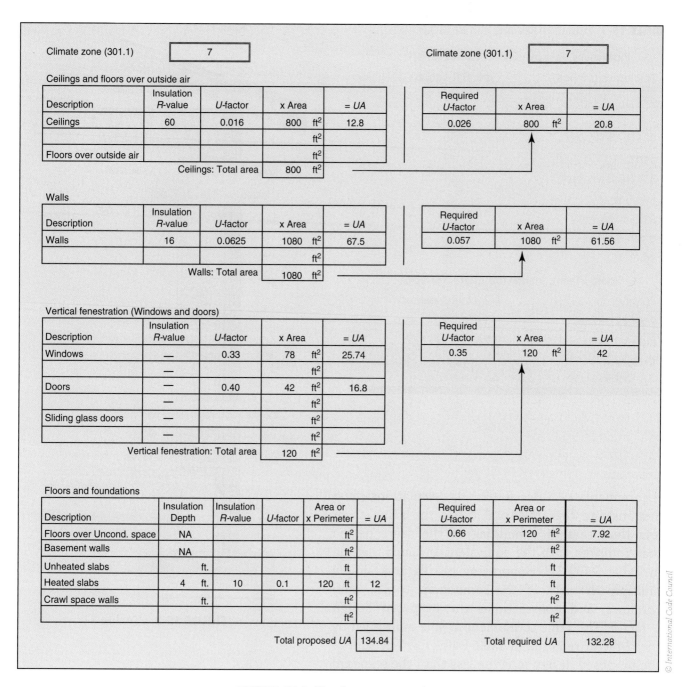

The table contained in the figure:

Climate zone (301.1): 7   Climate zone (301.1): 7

Ceilings and floors over outside air

Description	Insulation R-value	U-factor	x Area	= UA		Required U-factor	x Area	= UA
Ceilings	60	0.016	800 ft²	12.8		0.026	800 ft²	20.8
			ft²					
Floors over outside air			ft²					
Ceilings: Total area			800 ft²					

Walls

Description	Insulation R-value	U-factor	x Area	= UA		Required U-factor	x Area	= UA
Walls	16	0.0625	1080 ft²	67.5		0.057	1080 ft²	61.56
			ft²					
Walls: Total area			1080 ft²					

Vertical fenestration (Windows and doors)

Description	Insulation R-value	U-factor	x Area	= UA		Required U-factor	x Area	= UA
Windows	—	0.33	78 ft²	25.74		0.35	120 ft²	42
	—		ft²					
Doors	—	0.40	42 ft²	16.8				
	—		ft²					
Sliding glass doors	—		ft²					
	—		ft²					
Vertical fenestration: Total area			120 ft²					

Floors and foundations

Description	Insulation Depth	Insulation R-value	U-factor	Area or x Perimeter	= UA		Required U-factor	Area or x Perimeter	= UA
Floors over Uncond. space	NA			ft²			0.66	120 ft²	7.92
Basement walls	NA			ft²				ft²	
Unheated slabs	ft.			ft				ft	
Heated slabs	4 ft.	10	0.1	120 ft	12			ft	
Crawl space walls	ft.			ft²				ft²	
				ft²				ft²	

Total proposed UA: 134.84   Total required UA: 132.28

FIGURE 15-4 U-value computation

You Should Know

The 2009 American Society of Heating, Refrigerating and Air-Conditioning Engineers, Inc. (ASHRAE) *Handbook of Fundamentals* is a standard referenced in Chapter 5 of the IECC. **[Ref. R402.1.4 and Table 405.5.2(1)]** The handbook covers basic principles and includes data to support complete simple and complex calculations to demonstrate energy code compliance. Chapter 26, "Heat, Air, and Moisture Control in Building Assemblies—Material Properties," includes R-values for common building and insulating materials in Table 4. Values referenced in this table are used to determine equivalent U-factor or total UA (thermal transmittance) alternative compliance. ●

RES*check*™ is a free software program available from the U.S. Department of Energy. It can be used to evaluate performance design to show compliance with the provisions of Section R405. ResCheck can be used to demonstrate that the proposed building meets the efficiency requirements of the code by choosing "Performance Alternative" in the Compliance Method menu. Compliance with R405 does not eliminate the mandatory requirements that are referenced in R401.2. As an example, the air leakage provisions of Section R402.4 are still applicable even though the software calculations or performance alternative may indicate that the building is energy efficient. This is important information and is now easy to find in the code book because the mandatory requirements are marked as a reminder.

CEILINGS AND ATTICS

The IECC introduces two exceptions to Table R402.1.1 as specific insulation requirements. When framing or truss design allows the ceiling insulation to extend over the top of the exterior wall plate, the insulation R-value can be decreased from R-38 to R-30 or from R-49 to R-38. The truss design is specified and commonly called out as a "raised-heel" or "energy truss." Another exception is offered for cathedral ceilings or ceilings without attic spaces. This construction technique creates bigger-feeling interior spaces by applying a gypsum wall board (GWB) ceiling finish directly to the underside of the sloped roof rafter. This limits the available insulation to the depth of the roof or ceiling framing member. In climate zones where ceiling insulation is required to be R-38 or R-49, not more than 500 square feet of the ceiling area can be insulated to R-30 when the framing member depth will not accommodate the insulation required by Table R402.1.1. [Ref. R402.2.2]

These exceptions are good examples of apparent code requirement conflicts. The ceiling insulation requirements are reduced by specific conditions. Although Table R402.1.1 specifies an insulation level, the ceiling insulation requirements complying with these specific conditions are reduced. Section R101.4 clarifies this: "Where there is a conflict between a general requirement and a specific requirement, the specific requirement shall govern." The "raised heel" or "energy truss" R-value reduction allowance is specific to the insulation requirement in the table. This code provision is not to be confused with the other sentence in R101.4: "Where, in any specific case, different sections of this code specify different materials, methods of construction or other requirements, the most restrictive shall govern." Because the raised heel and cathedral ceiling provisions are a specific situation identified in the construction document details, these take precedence over the generally required insulation provisions from Table 402.1.1.

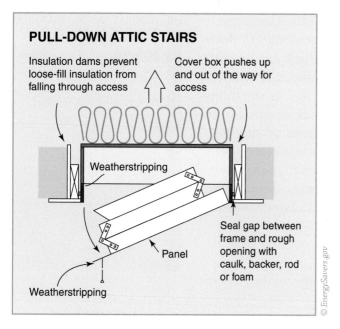

PULL-DOWN ATTIC STAIRS

Insulation dams prevent loose-fill insulation from falling through access

Cover box pushes up and out of the way for access

Weatherstripping

Seal gap between frame and rough opening with caulk, backer, rod or foam

Panel

Weatherstripping

© EnergySavers.gov

FIGURE 15-5 Attic pull-down stairs

Attic Ventilation

Attic ventilation is required by Section 1203.2 of the *International Building Code* (IBC) and Section R806 of the *International Residential Code* (IRC). The insulation must be held back and not block the vent area. A baffle needs to be installed next to eave or soffit vents and maintain an open space the same size as or larger than the size of the vent. **[Ref. R402.2.3]** This detail should be included in the plans at permit submittal. IRC 806.4 includes provisions for unvented attic or rafter spaces.

Attic access is required by IBC 1209.2 and IRC R807. The hatches or doors that connect conditioned and unconditioned spaces must be insulated to the same or greater R-value as the area around them. **[Ref. 402.2.4]** This maintains the integrity of the thermal envelope. These are often site built and do not require special methods or materials (Figure 15-5).

The IRC and IBC do not reference the energy code in these respective sections; however, the provisions for eave baffles and attics and crawlspaces must be included in the building plan documents and details.

WALLS

The energy code recognizes no fewer than seven different types of walls in Section R402.2. This is a necessary distinction because each material in the assembly behaves differently in the thermal envelope and because some walls are above grade, below grade (Figure 15-6), or both. A typical walkout basement or garden-level apartment has wood or steel-frame walls, basement walls, and perhaps a crawlspace wall. The insulation values and often the insulation type and location are different for each. This makes proper identification of each wall type extremely important, and the code relies on designers, builders, plan examiners, and building inspectors to understand the definitions and distinctions.

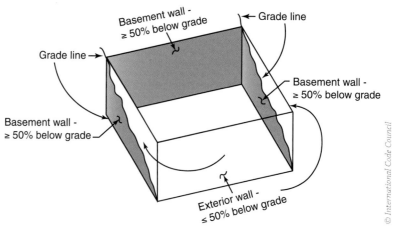

Basement wall - ≥ 50% below grade

Grade line

Grade line

Basement wall - ≥ 50% below grade

Basement wall - ≥ 50% below grade

Exterior wall - ≤ 50% below grade

© International Code Council

FIGURE 15-6 Typical above-/below-grade wall construction

The definitions for the various wall types are scattered around Chapter 2 of the IECC and in no particular order. Remember that when a word or term in the code is defined in Chapter 2, it is *italicized*. In alphabetical order, *above-grade wall*, *basement wall*, *crawl space wall* (do not get confused with *curtain wall*), and *exterior wall* are specifically defined. This leaves a few types of walls undefined, but careful reading of the definitions can establish the proper application of the appropriate energy code requirement. [Ref. R202]

Code Basics

Wall construction on a sloping grade typically requires the below-grade portions to use materials designed for basement wall construction and the above-grade portions to use materials for exterior walls. The "stepped wall" must be evaluated as either a basement or exterior wall by measuring the percentage of the wall below grade; 50 percent or more wall surface below grade defines a basement wall. Each part of the wall must meet the insulation requirements. The same basement wall provisions apply to the above-grade wall construction even if it is built with materials typical of exterior wall construction. ●

Mass Walls

Mass walls are built of dense materials such as concrete block, reinforced concrete (Figure 15-7), brick, and insulated concrete forms (ICFs). Traditional building materials and methods such as adobe (Figure 15-8), solid logs, and rammed earth are also mass wall materials. Frame walls and mass walls are very different in both performance and the amount of insulation required. Footnotes to both prescriptive tables require insulation values that are dependent on the location of the insulation (Table 15-2). It makes a difference as to whether more than half of the insulation is on the interior of the mass wall. In all climate zones,

© photobank.ch/www.shutterstock.com

FIGURE 15-7 Concrete mass wall

FIGURE 15-8 Adobe wall

TABLE 15-2 Mass wall table

Climate Zone	Mass Wall *R*-Value[i]
1	3/4
2	4/6
3	8/13
4 except Marine	8/13
5 and Marine 4	13/17
6	15/20
7 and 8	19/21

[i] The second R-value applies when more than half the insulation value is on the interior of the mass wall

TABLE R402.1.3 Equivalent *U*-Factors

Climate Zone	Mass Wall *U*-Factors[c]
1	0.197
2	0.165
3	0.098
4 except Marine	0.098
5 and Marine 4	0.082
6	0.060
7 and 8	0.057

[c] Basement wall U-Factor of 0.360 in warm-humid locations as defined by figure R301.1 and Table R301.1.

more insulation is required when more than half of the insulation is on the interior of the mass wall. [Ref. R402.2.5]

Steel-Frame Walls

Steel framing is sometimes specified and is common in multi-family buildings (Figure 15-9); it must comply with the requirements of the residential energy code provisions. [Ref. R402.2.6] Steel has a much higher thermal conductivity than wood—think about how much colder a steel column feels in the winter than a wood column. To adjust for the thermal transfer properties of steel, Table R402.2.6 (see Table 15-3)

FIGURE 15-9 Steel framing

TABLE 15-3 Steel Frame Insulation

TABLE R402.2.6 Steel-Frame Ceiling, Wall, and Floor Insulation (*R*-Value)

Wood Frame *R*-Value Requirement	Cold-Formed Steel Equivalent *R*-Value[a]
Steel Truss Ceilings[b]	
R-30	R-38 or R-30 + 3 or R-26 + 5
R-38	R-49 or R-38 + 3
R-49	R-38 + 5
Steel Joist Ceilings[b]	
R-30	R-38 in 2 × 4 or 2 × 6 or 2 × 8 R-49 in any framing
R-38	R-49 in 2 × 4 or 2 × 6 or 2 × 8 or 2 × 10
Steel-Framed Wall 16" O.C.	
R-13	R-13 + 4.2 or R-19 + 2.1 or R-21 + 2.8 or R-0 + 9.3 or R-15 + 3.8 or R-21 + 3.1
R-13 + 3	R-0 + 11.2 or R-13 + 6.1 or R-15 + 5.7 or R-19 + 5.0 or R-21 + 4.7
R-20	R-0 + 14.0 or R-13 + 8.9 or R-15 + 8.5 or R-19 + 7.8 or R-19 + 6.2 or R-21 + 7.5
R-20 + 5	R-13 + 12.7 or R-15 + 12.3 or R-19 + 11.6 or R-21 + 11.3 or R-25 + 10.9
R-21	R-0 + 14.6 or R-13 + 9.5 or R-15 + 9.1 or R-19 + 8.4 or R-21 + 8.1 or R-25 + 7.7
Steel Framed Wall, 24" O.C.	
R-13	R-0 + 9.3 or R-13 + 3.0 or R-15 + 2.4
R-13 + 3	R-0 + 11.2 or R-13 + 4.9 or R-15 + 4.3 or R-19 + 3.5 or R-21 + 3.1
R-20	R-0 + 14.0 or R-13 + 7.7 or R-15 + 7.1 or R-19 + 6.3 or R-21 + 5.9
R-20 + 5	R-13 + 11.5 or R-15 + 10.9 or R-19 + 10.1 or R-21 + 9.7 or R-25 + 9.1
R-21	R-0 + 14.6 or R-13 + 8.3 or R-15 + 7.7 or R-19 + 6.9 or R-21 + 6.5 or R-25 + 5.9
Steel Joist Floor	
R-13	R-19 in 2 × 6, or R-19 + 6 in 2 × 8 or 2 × 10
R-19	R-19 + 6 in 2 × 6, or R-19 + 12 in 2 × 8 or 2 × 10

[a] Cavity insulation *R*-value is listed first, followed by continuous insulation *R*-value.

[b] Insulation exceeding the height of the framing shall cover the framing.

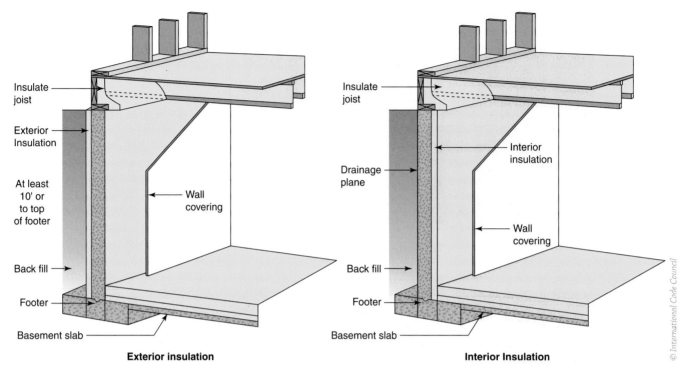

FIGURE 15-10 Basement wall insulation

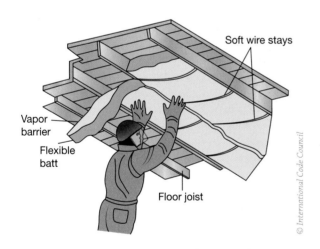

FIGURE 15-11 Basement ceiling insulation

offers steel-frame insulation combinations equivalent to the wood-frame R-value requirements. Combinations of cavity and continuous insulation offer cost and constructability options for the designer and builder choosing steel framing in residential buildings.

Basement Walls

In conditioned basements, walls (see definitions of *basement wall* and *above-grade wall*) **[Ref. R202]** must be insulated from the top of the basement wall to at least 10 feet below grade or to the basement floor, whichever is less (Figure 15-10). It is very important for the designer and builder to clearly indicate the final grade on the plans. This will speed building department review for compliance and permit issuance. If the basement is designed as unconditioned space, the thermal envelope boundary must be shifted to the underside of the floor above (Figure 15-11). In this condition the rim joist is considered to be part of the thermal envelope, and care must be taken to insulate it to the same requirements as for the basement walls. **[Ref. R402.2.8]**

Crawlspace Walls

Crawlspaces must be insulated. The code allows for the thermal envelope to be horizontal in the floor above or vertical to create a conditioned crawlspace. Unvented or vertically insulated crawlspaces are allowed

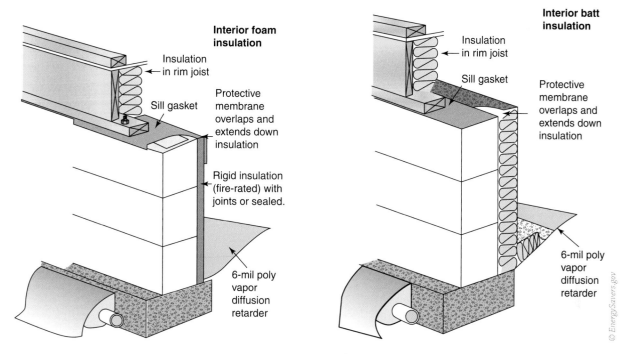

FIGURE 15-12 Interior insulated crawlspace

in IRC R408.3 and IBC 1203.3.2 exception 4. Unvented crawlspaces are popular in colder climates to prevent the freezing of pipes. In all climates, a Class I vapor retarder, typically sheet polyethene, must cover the exposed dirt in the crawlspace. The horizontal overlap joints in the vapor retarder membrane must be taped or sealed and the retarder must extend vertically 6 inches up and be sealed and attached to the stem wall (Figure 15-12).

The crawlspace insulation must cover the entire wall from the floor above to the grade below and then extend horizontally or vertically at least 24 inches down the stem wall. The insulation below the crawlspace floor next to the stem wall partially eliminates the "thermal bridge" created at the intersection of the foundation wall and crawlspace floor (Figure 15-13). **[Ref. R402.2.10]**

FLOORS

Insulation requirements for wood-frame floor joist construction are easy to find in the now familiar R-value and U-value tables in Section R402.

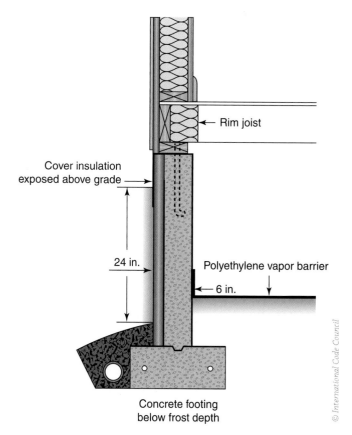

FIGURE 15-13 Exterior insulated crawlspace

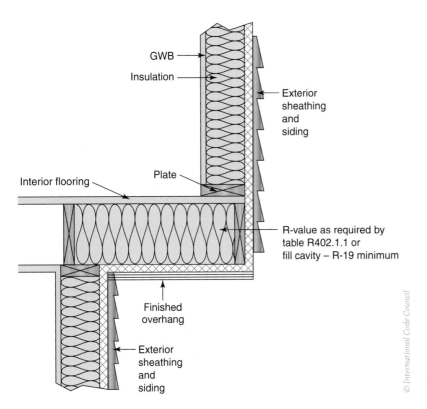

GWB

Insulation

Exterior
sheathing
and
siding

Interior flooring

Plate

R-value as required by
table R402.1.1 or
fill cavity – R-19 minimum

Finished
overhang

Exterior
sheathing
and
siding

© International Code Council

FIGURE 15-14 Cantilever floor insulation

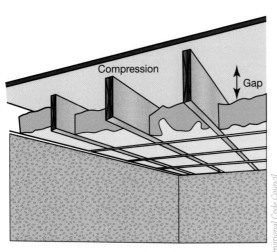

Compression

Gap

Two common flaws in floor insulation are gaps above
the batt and compression of the batt in the cavity.

© International Code Council

FIGURE 15-15 Sagging floor insulation

The table includes an important footnote g that is an exception to the minimum insulation requirements. In climate zones where R-30 or R-38 insulation is required, R-19 is an acceptable minimum. The framing cavity must be filled and depending on the depth of the floor joists, the R-value may be greater than R-19. This footnote acknowledges that increasing joist size to accommodate R-30 or R-38 insulation adds considerable cost to construction and is not justified in energy savings over the useful life of the building.

Floors framed with steel joists are required to meet the equivalent R-values listed in Table R402.2.6. Notice that in all but one application, continuous insulation is required to comply with the steel-frame table for floors. The constructability of this assembly must be carefully considered during the design phase of plan development. Sufficient detail must be included in the plan documents to demonstrate compliance in the permit review and field inspection approvals.

In both wood- and steel-frame floor assemblies the insulation must be installed to maintain permanent contact with the underside of the subfloor above (Figure 15-14). If the insulation sags or droops, cold air will move between the insulation and floor, minimizing the effectiveness of the insulation and creating uncomfortable cold spots in the floor (Figure 15-15). **[Ref. R402.2.7]**

The R-value, U-value, and installation requirements apply to all floors that are part of the thermal envelope and do not apply to floors with conditioned space both above and below the floor.

Slab-on-Grade Floors

Many buildings are designed with slab-on-grade floors. This technique saves the cost of excavating the site to accommodate a basement or crawlspace. It is also possible to incorporate the slab-on-grade in portions of the design and crawlspace or basement construction in the same building. Building site slope and topography usually dictate the decisions designers and builders make to maximize the building's integration with its surroundings.

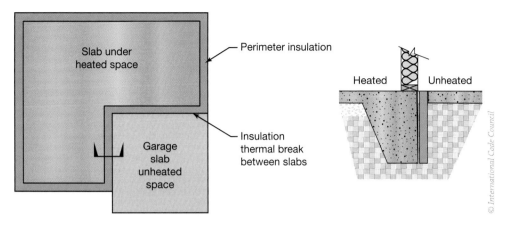

FIGURE 15-16 Mixed slab

Any portion of the conditioned space that incorporates a slab 12 inches or less below grade requires insulation. The insulation creates a thermal break between the interior conditioned space and earth adjacent to the concrete footing and slab. Heated slabs require additional R-5 insulation at the slab edges. **[Ref. R402.2.9]** The U.S. Department of Energy estimates that slab edge insulation can reduce winter heating bills by 10 to 20 percent, and in new construction the annual energy savings exceed the added cost of construction when factored into the mortgage cost. The investment in slab insulation pays off from the beginning.

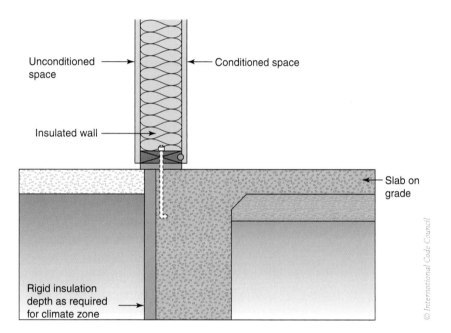

FIGURE 15-17 Insulation detail

Unheated slabs in climate zones 1, 2, and 3 do not require insulation. Notice that the U-value Table R402.1.3 does not offer an alternative to the R-value and depth requirements in Table R402.1.1. Slab-on-grade design serving conditioned spaces must meet the minimums indicated in Table 402.1.1 (see Figure 15-16). Footnote d is important, as it notes that an additional R-5—for a total of R-15 (10 + 5)—is required for all slab edges of heated slabs, except in climate zones 1, 2, and 3, where the R-5 additional insulation is adequate. The insulation depth of 2 or 4 feet is determined by climate zone (Figure 15-17). **[Ref. Table 402.1.1]**

The only exception allowed for slab-on-grade insulation is in localities with a very heavy infestation of termites. Termite infestation probability maps are found in IRC Figure R301.2 (6) and IBC Figure 2603.9. However, every building mapped in the "very heavy" areas is not automatically exempt from the insulation requirements. It is very important

FIGURE 15-18 Thermally isolated sunroom

FIGURE 15-19 Non-thermally isolated sunroom

to check with the *code official* in the jurisdiction where the proposed building will be constructed.

SUNROOMS

Sunrooms are a special case in the energy code. These special-use rooms are defined as "a one-story structure attached to a dwelling with a glazing area in excess of 40 percent of the gross area of the structure's exterior walls and roof." The driving statement in the code requires all conditioned sunrooms to meet the insulation requirements of the code. The key design elements for conditioned sunrooms are found in the exceptions for sunrooms with *thermal isolation*. A thermally isolated sunroom is physically separated from the dwelling and the heating and cooling is controlled by a separate zone thermostat or by separate equipment (Figure 15-18). The minimum ceiling insulation requirements are reduced according to climate zones. The wall insulation requirement is R-13 in all climate zones. If the sunroom does not meet the definition of *thermal isolation* (Figure 15-19), the wall and ceiling insulation requirements are as required by the R-value or U-factor tables. [**Ref. R402.2.12**]

The windows, doors, and skylights in sunrooms must meet the minimum requirements for the applicable climate zone except when the attached space is thermally isolated. The U-factor for fenestration is relaxed for thermally isolated sunrooms in climate zones 4 through 8. R402.3.5 allows for vertical fenestration to have a maximum U-factor of 0.45 and a 0.70 U-factor for skylights.

FENESTRATION

The *Oxford Dictionary* defines *fenestration* in architecture as "the arrangement of windows in a building." The term *fenestration* in the

TABLE 15-4 Fenestration requirements by climate zone

TABLE R402.1.1 Insulation and Fenestration Requirements by Component

Climate Zone	Fenestration U-Factor[b]	Skylight U-Factor	Glazed Fenestration Shgc[b,c]
1	NR	0.75	0.25
2	0.40	0.65	0.25
3	0.35	0.55	0.25
4 except Marine	0.35	0.55	0.40
5 and Marine 4	0.32	0.55	NR
6	0.32	0.55	NR
7 and 8	0.32	0.55	NR

[b] The fenestration U-factor column excludes skylights. The SHGC column applies to all glazed fenestration. Exception: Skylights may be excluded from glazed fenestration SHGC requirements in Climate Zones 1 through 3 where the SHGC for such skylights does not exceed 0.30.

[c] There are no SHGC requirements in the Marine Zone.

energy code refers to light-transmitting areas in a building's thermal envelope. The definition includes windows, skylights, and opaque and glazed doors. The insulating value of a window is much lower than that of the wall it is installed in. *Fenestration* performance affects occupant comfort and building energy use.

Windows

Table R402.1.1 illustrates the relationship of window and skylight energy-saving performance characteristics and climate zones (see Table 15-4). Notice that the U-factor requirements range from "not required" (NR) in semi-tropical climates to 0.32 in semi-arctic climates. The required insulating quality of windows and skylights becomes increasingly more restrictive as the Heating Degree Days (HDD) become greater. Conversely, the SHGCs begin at 0.25 in the warmest climate zones, where the Cooling Degree Days (CDD) dominate the energy-use load and nearly every building is air conditioned. Solar heat gain can be a good thing in cooler climates and can be maximized with thoughtful and proper design. Table 402.1.1 reflects that wisdom in the three climate zones with no requirement for SHGC performance.

The U-factor and SHGC rating of windows, doors, and skylights is determined according to the standards referenced in R303.1.3. Every product is tested and labeled with the U-factor, SHGC, visible transmittance (VT), and air leakage values. If the fenestration product is not tested and labeled, the default tables in R303.1.3 must be used to determine compliance.

Not every window or skylight has to comply with U-factor and SHGC maximum and minimum requirements. The area-weighted-average U-factor approach offers an alternative path to compliance.

You Should Know

The National Fenestration Rating Council® is a nonprofit organization providing a labeling system for the energy performance of windows, doors, and skylights. The ratings establish a uniform method to compare fenestration products and design benefits desirable for climate, location, and building orientation.

The numbers in the boxes on the label use the terms defined in the energy code. **[Ref. R202]** Find these terms in the code book, and then compare the code's definitions to the following information from the U.S. Environmental Protection Agency (EPA) Energy Star publication:

- **U-Factor** measures the rate of heat transfer and tells you how well the window insulates. U-factor values generally range from 0.25 to 1.25 and are measured in Btu/h·ft²·°F. The lower the U-factor, the better the window insulates.
- **Solar Heat Gain Coefficient (SHGC)** measures the fraction of solar energy transmitted and tells you how well the product blocks heat caused by sunlight. SHGC is measured on a scale of 0 to 1; values typically range from 0.25 to 0.80. The lower the SHGC, the less solar heat the window transmits.
- **Visible Transmittance (VT)** measures the amount of light the window lets through. VT is measured on a scale of 0 to 1; values generally range from 0.20 to 0.80. The higher the VT, the more light you see.
- **Air Leakage (AL)** measures the rate at which air passes through joints in the window. AL is measured in cubic feet of air passing through one square foot of window area per minute. The lower the AL value, the less air leakage. Most industry standards and building codes require an AL of 0.3 cf·m/ft².
- **Condensation Resistance** measures how well the window resists water build-up. Condensation Resistance is scored on a scale from 0 to 100. The higher the Condensation Resistance factor, the less buildup the window allows. ●

[Ref. R402.3.1] The concept of averaging glazing performance and the total square feet in the proposed design is straightforward and easily illustrated in a simple example:

Example:

These window types are included in the schedule for a single-family home proposed in climate zone 7. The U-factor is noted on the window schedule.

Window type 1	U-0.30	125 ft²
Window type 2	U-0.32	75 ft²
Window type 3	U-0.35	25 ft²

$$\text{Answer:} \quad \frac{(125 \text{ ft}^2 \times .30) + (75 \text{ ft}^2 \times .32) + (25 \text{ ft}^2 \times .35)}{225 \text{ ft}^2} = U - 0.312 \text{ average}$$

U-0.32 is required in climate zone 7, and 0.312 is less than 0.32, so the proposed windows complies using a weighted-average calculation.

The area-weighted compliance approach to SHGC is similar to but not exactly like the U-factor method. Only fenestration that is more than 50 percent glazed is to be included in the calculation. Entry doors, for example, may be less than half glass, whereas patio doors may be mostly glass.

This makes sense, as the heat gain potential is a direct function of the amount of glazing in the fenestration product. [Ref. R402.3.2]

Example:

These window types are included in the schedule for a single-family home proposed in climate zone 3. The SHGC value is noted on the window schedule.

Window type 1	SHGC-.23	125 ft²
Window type 2	SHGC-.25	75 ft²
Window type 3	SHGC-.28	25 ft²

$$Answer: \frac{(125 \text{ ft}^2 \times .23) + (75 \text{ ft}^2 \times .25) + (25 \text{ ft}^2 \times .28)}{225 \text{ ft}^2} = \text{SHGC .242 average}$$

SGHC .25 is required in climate zone 3, and .242 is less than 0.25, so the proposed windows complies using a weighted-average calculation.

The code does place limits on the performance of fenestration allowed in the area-weighted calculation method to demonstrate compliance. The mandatory provisions in R402.5 list maximum U-factors and SHGC values for vertical fenestration and skylights by climate zone. This code section eliminates the use of any fenestration product that exceeds these maximums in the two tradeoff methods to demonstrate compliance. Fenestration U-factors and SHGC values in designs using the simulated performance method are limited by the restrictions in this section. [Ref. R402.5]

Every door or window is required to be labeled with the NFRC 100 test results. The field inspector will check each unit for compliance with the window and door schedule. If the product is not tested, the values in the three default tables will determine the rating (see Tables 15-5, 15-6, 15-7). It is important

TABLE 15-5 TABLE R303.1.3(1) Default Glazed Fenestration *U*-Factor

Frame Type	Single Pane	Double Pane	Skylight Single	Skylight Double
Metal	1.20	0.80	2.00	1. 30
Metal with Thermal Break	1.10	0.65	1.90	1.10
Nonmetal or Metal Clad	0.95	0.55	1.75	1.05
Glazed Block		0.60		

© International Code Council

TABLE 15-6 TABLE R303.1.3(2) Default Door *U*-Factors

Door Type	*U*-Factor
Uninsulated Metal	1.20
Insulated Metal	0.60
Wood	0.50
Insulated, nonmetal edge, max 45 percent glazing, any glazing double pane	0.35

© International Code Council

TABLE 15-7 TABLE R303.1.3(3) Default Glazed Fenestration SHGC and VT

	Single Glazed		Double Glazed		Glazed Block
	Clear	Tinted	Clear	Tinted	
SHGC	0.8	0.7	0.7	0.6	0.6
VT	0.6	0.3	0.6	0.3	0.6

to note that these default values exceed the maximum values allowed to be used in the area-weighted calculation.

Doors

The energy code offers two very practical exemptions to all of the fenestration requirements. The intent is to allow some design flexibility for designers and contractors to include decorative windows, skylights, and glazed doors. Up 15 square feet of glazing per dwelling is exempt from both the U-factor and SHGC code requirements, but there is a condition: The exemption can only be taken using the fenestration component requirements specified in Table R402.1.1. A similar exemption is available to exclude a side-hinged opaque door assembly no bigger than 24 square feet. Again, it is important to understand that the design does not qualify for these two exemptions if the U-factor or total UA alternative paths are used to demonstrate energy code compliance. **[Ref. R402.3.4]**

Fenestration Replacement

In work that requires removing a window sash, a skylight window frame, or glazed door, the new fenestration components are required to comply with the U-factors and SHGC values for new construction. This is an important distinction when considering Exception 2 in R101.4.3 that allows any type of glass pane to be installed in an existing sash. When the sash and/or frame are replaced, the new units must meet the applicable requirements in Table R402.1.1 (Figure 15-20). When only a window pane is replaced, the glazing is not required to comply (Figure 15-21).

Infiltration

Air leakage in the the building envelope is called infiltration even though it is a combination of infiltration and exfiltration. Air moves into and out of the the conditioned space when the pressure inside the thermal envelope is less than or greater than the air pressure outside. Exfiltration describes air movement through the building thermal envelope to the outside. All of these examples contribute air movement through the building thermal envelope. This condition allows conditioned air to escape and creates a particularly inefficient building energy use. Infiltration and exfiltration occur at the same time while the building works

FIGURE 15-20 Window replacement

to balance itself, with air finding cracks and holes in the *air barrier* to move in and out.

Many building designs rely on natural ventilation, meaning an open or closed window, to provide critical building pressure control. Other design strategies incorporate a balanced controlled ventilation system. Heat recovery ventilation (HRV) and energy recovery ventilation (ERV) supply fresh air to the building interior and "recover" energy from the tempered exhaust air. HRV and ERV systems prove to be very efficient and effective in buildings with very little air leakage.

FIGURE 15-21 Window pane replacement glazing is not regulated by IECC

The 2012 IECC recognizes that air leakage control is absolutely critical in the "construction of buildings for the effective use and conservation of energy over the useful life of each building." To achieve the stated intent of the energy code, new provisions, including mandatory testing, are now requirements. [Ref. R402.4]

THE AIR BARRIER AND BUILDING THERMAL ENVELOPE

The specific requirement for the *continuous air barrier* is that it must be installed in the building thermal envelope. An easy test to check for compliance is to trace the air barrier location in the building section details on the construction documents. If this can be done without lifting the pencil off the paper, the air barrier is continuous. Table R402.4.1.1 lists 16 building envelope elements that require special attention to detail. Construction document notes and callouts will address how the building construction complies with these requirements. Table 15-8 highlights items to be detailed on the construction documents.

Caulk and sealant materials and methods in air-barrier construction must allow for thermal and moisture expansion and contraction. Joints between different materials have to shrink and stretch as the air-barrier components go through daily heat-up and evening cool-down cycles. Any material used must be installed per the manufacturer's installation instructions. The instructions must be on site and available for review during the field inspection process. [Ref. R402.4.1]

Fireplaces and recessed luminaires (light fixtures) merit special consideration in air-barrier detailing and inspection. New wood-burning fireplaces must have tight-fitting flue dampers and outdoor supplied combustion air. [Ref. R402.4.2] The damper requirement keeps conditioned air from flowing out the chimney, and the outdoor supplied combustion air requirement ensures the fireplace pulls in unconditioned outside air to burn instead of pulling air from inside the home that has already been conditioned. Fireplace construction requirements are also found in IRC Chapter 10 and IBC Chapter 2111.

TABLE 15-8 Table R402.4.1.1 (2012 IECC). Air Barrier and Insulation Installation

Component	Criteria*
Air barrier and thermal barrier	A continuous air barrier shall be installed in the building envelope. Exterior thermal envelope contains a continuous air barrier. Breaks or joints in the air barrier shall be sealed. Air-permeable insulation shall not be used as a sealing material.
Ceiling/attic	The air barrier in any dropped ceiling/soffit shall be aligned with the insulation and any gaps in the air barrier sealed. Access openings, drop-down stair or knee wall doors to unconditioned attic spaces shall be sealed.
Walls	Corners and headers shall be insulated and the junction of the foundation and sill plate shall be sealed. The junction of the top plate and top of exterior walls shall be sealed. Exterior thermal envelope insulation for framed walls shall be installed in substantial contact and continuous alignment with the air barrier. Knee walls shall be sealed.
Windows, skylights and doors	The space between window/door jambs and framing and skylights and framing shall be sealed.
Rim joists	Rim joists shall be insulated and include the air barrier.
Floors (including above-garage and cantilevered floors)	Insulation shall be installed to maintain permanent contact with underside of subfloor decking. The air barrier shall be installed at any exposed edge of insulation.
Crawl space walls	Where provided in lieu of floor insulation, insulation shall be permanently attached to the crawl space walls. Exposed earth in unvented crawl spaces shall be covered with a Class 1 vapor retarder with overlapping joints taped.
Shafts, penetration	Duct shafts, utility penetrations, and flue shafts opening to exterior or unconditioned space shall be sealed.
Narrow cavities	Batts in narrow cavities shall be cut to fit, or narrow cavities shall be filled by insulation that on installation readily conforms to the available cavity space.
Garage separation	Air sealing shall be provided between the garage and conditioned spaces.
Recessed lighting	Recessed light fixtures installed in the building thermal envelope shall be air tight, IC rated, and sealed to the drywall.
Plumbing and wiring	Batt insulation shall be cut neatly to fit around wiring and plumbing in exterior walls, or insulation that on installation readily conforms to available space shall extend behind piping and wiring.
Shower/tub on exterior wall	Exterior walls adjacent to showers and tubs shall be insulated and the air barrier installed separating them from the showers and tubs.
Electrical/phone box on exterior walls	The air barrier shall be installed behind electrical or communication boxes or air sealed boxes shall be installed.
HVAC register boots	HVAC register boots that penetrate building thermal envelope shall be sealed to the subfloor or drywall.
Fireplace	An air barrier shall be installed on fireplace walls. Fireplaces shall have gasketed doors.

• In addition, Inspection of log walls shall be in accordance with the provisions of ICC-400.

Recessed light fixtures installed in the thermal envelope must be air sealed. A gasket or caulk between the recessed light housing and the interior wall or ceiling is required. Because the lights are penetrating the thermal envelope, the units must be IC-rated (insulation contact) and tested and labeled for air leakage. A sealed box over the can light fixture with proper detailing also complies. The box must be sealed to the attic floor, the wire penetrations into the box must be sealed, and the entire assembly must be installed with a minimum 3-inch clearance to adjacent insulation. [Ref. R402.4.4]

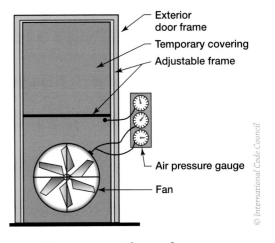

FIGURE 15-22 Blower door test

BLOWER DOOR TEST

A blower door test is mandatory to verify that the requirements for maximum air leakage are met. The calibrated blower door assembly is elegant and simple; an easily portable instrument fits into an exterior door frame. A powerful variable-speed fan pulls air out of the building or dwelling being tested, and outside air flows inside through unsealed cracks and holes (Figure 15-22). The higher the fan speed, the less effective the air barrier. Specific test criteria must be followed to maintain comparable and consistent measurement results. Preparation and criteria for the test and required performance are addressed in R402.4.1.2.

The blower door test must confirm that the total air leakage is less than 5 air changes per hour (ACH) in climate zones 1 and 2 and not more than 3 ACH in climate zones 3 through 8. That means that every 12 minutes a dwelling in climate zone 1 or 2 gets a complete interior air change, and every 20 minutes in the other climate zones. The test results must be submitted to the code official in a written, signed report. [Ref. R02.4.1.2]

You Should Know

Unlike previous editions of the IECC, in the 2012 edition there is no option for testing *or* visual inspection. The requirement is now testing *and* visual inspection. The code official is authorized to require an approved third party to inspect and verify. ●

You Should Know

Blaise Pascal was a hardworking French philosopher, mathematician, and physicist in the mid-seventeenth century. In his short 39-year life he developed a mathematical probability theory used to make gambling and economics more or less predictable, invented the first mechanical calculator, and may have worn one of the first wristwatches.[1] Pascal is obviously the namesake for the pascal (Pa) unit of measurement, a unit of atmospheric pressure equal to 1 newton per square meter, in the International System of Units (SI). Standard atmospheric pressure is 101,325 Pa.

Blower door tests are calibrated to a 50-Pa pressure difference between the inside and outside of the building air barrier. Fifty Pa of pressure is equal to .00725 psi, which is not very much. One pascal of pressure is about the weight of one postage stamp. ●

[1] William J. Lynott, *Watch Development*, February 2012.

SECTION 16

Building Systems

© istockphoto/mustafa deliormanli

Mechanical ventilation is a mandatory requirement in the 2012 *International Energy Conservation Code* (IECC). The requirement is referenced in Section R303.4 of the *International Residential Code* (IRC) and Section 401.3 of the *International Mechanical Code* (IMC). Both codes state that mechanical ventilation is required when the air infiltration rate in the dwelling is less than 5 ACH. These code provisions align requirements regarding mechanical ventilation. Every mechanical system must be designed with controls, delivery mechanisms, and properly sized equipment. Specific provisions that support energy savings in HVAC and hot water systems are also included.

PROGRAMMABLE THERMOSTAT

Adjusting the demand for heating and cooling the dwelling saves energy. All forced-air furnace heating and cooling systems must have at least one programmable thermostat. It must be capable of daily time and temperature settings. Settings for *zones* range from 55°F to 85°F. This required feature may be used as a "vacation" from typical heating and cooling requirements and can compound energy savings when the dwelling is unoccupied. The programmable thermostat must be specified to limit the heating temperature to no higher than 70°F and the cooling temperature to no lower than 78°F, and must be referenced and called out in the plan documentation. This is the initial setting that is verified and approved at the final building inspection.

DUCTS

Ducts are the delivery passageways for the ventilation system that moves fresh air into and stale air out of the dwelling. The effectiveness of the ductwork is critical to efficient energy use in the building and providing comfort for the people living in the building. Ducts must transfer conditioned air under positive pressure from the air-handling unit to rooms around the unit and take stale air out under negative pressure for recirculation or exhaust. Provisions specific to ducts and air handlers improve energy efficiency in the design and installation of these systems.

Supply ducts must be insulated unless installed completely inside the building thermal envelope. R-8 insulation is required to protect the conditioned air moving through supply ducts in unconditioned attics. This helps prevent condensation (sweating) in warmer months on air-conditioner ducts and keep heated air warmer in the cooler months. Ductwork that is not completely inside the building thermal envelope must be installed with at least R-6 insulation.

Many HVAC energy experts estimate[1] that about 20 percent of conditioned air intended for distribution in the dwelling unit does not make it to the room or space due to leaks, holes, and poorly constructed ductwork systems. These losses result in higher energy bills and a lower level of occupant comfort. Ducts, air handlers, and filter boxes must be sealed and inspected during construction (Figure 16-1). The IMC and IRC give direction as to compliant materials, techniques, and standards regarding duct construction. These requirements apply to all ducts in all construction. The IECC requires that duct system air-tightness be tested and

> ## Code Basics
>
> Multiple system designs are possible for an air distribution system to deliver code-compliant energy efficiency and provide a comfortable thermal environment. Duct sizing is required by IMC Section 603.2 and references ACCA Manual D as an approved method. Duct sizing methods consider building floor plan, room sizes, number of levels, the size of windows in each room, and forced air furnace location. ACCA Manual D establishes design parameters and calculation methods to provide an air-handling system for comfortable air flows to each room. ●

FIGURE 16-1 Duct seal-mastic

© International Code Council

[1]www.energystar.gov (see Home Improvement—Duct Sealing).

verified at either the mechanical rough-in or final inspection. Ducts and air handlers located entirely within the building thermal envelope do not have to be duct-blaster tested.

Designers, builders, plans examiners, and building inspectors must be familiar with the requirements and specific exceptions applicable to mechanical system components intentionally located entirely within the building thermal envelope. Well-designed and properly sealed duct systems save energy and can make the dwelling more comfortable.

EXAMPLE

© International Code Council

FIGURE 16-2 Twisted duct

The duct construction in Figure 16-2 shows three Type 8AA elbow fittings used to achieve one 90-degree turn. How does this change the equivalent length in the duct flow calculation (Figure 16-3)?

		Elbows and Offsets	
Picture	ID	Equivalent length	Fitting description
	8A6	20	Smooth elbow, R/D = 0.75
	8A9	15	Smooth elbow, R/D = 1.0
	8AE	10	Smooth elbow, R/D = 1.5
	8AA	20	4 or 5 piece elbow, R/D = 1.0
	8AF	15	4 or 5 piece elbow, R/D = 1.5
	8A8	35	3 piece elbow, R/D = 0.75
	8AB	25	3 piece elbow, R/D = 1.0
	8AH	20	3 piece elbow, R/D = 1.5

Courtesy of ACCA

FIGURE 16-3 Manual D duct sizing

Answer: Each 8AA elbow adds 20 feet in equivalent duct length. The two additional elbows add another 40 feet. This will affect the static duct pressure and the amount of fan pressure needed in the duct to deliver the cubic feet per minute (CFM) in the room as required by ACCA Manual D.

HOT WATER SYSTEMS

Mechanical system pipes designed to carry hot water above 105°F or cold water less than 55°F are required to be insulated. R-3 pipe insulation must be installed and protected against damage. Equipment maintenance, sunlight, and moisture all pose a threat to the expected performance of pipe insulation over the useful life of the building.

FIGURE 16-4 Insulating hot water pipes reduces heat loss

Insulating hot water pipes reduces heat loss (Figure 16-4). Table R403.4.2 introduces a pipe diameter and maximum pipe run approach to determine the insulation requirement.

This concept is also a water-conserving measure because warmer water is delivered to the fixture without waiting for the cooler water to run down the drain. Table R403.4.2 basically lists exceptions to the nine conditions and all other piping requiring R-3 pipe insulation. Only hot water piping sizes shorter than the maximum lengths specified in Table 403.4.2 (see Table 16-1) and not covered by one of the nine other conditions are exempt from the R-3 pipe insulation requirement.

Notice that piping outside the conditioned space is required to be insulated. The energy code is consistent: Keeping building systems completely inside the building thermal envelope is a good design idea!

MECHANICAL VENTILATION

Mechanical ventilation systems are covered by requirements in the 2012 IECC. The designs are either exhaust-only, supply-only, or balanced air-exchange systems (Table 16-2). System details regarding sources of supply air, locations of exhaust air, and ventilation rates are provided in IRC M1507.3 and IMC Section 403. The IMC states that "the amount of supply air shall be approximately equal to the amount of return and exhaust air," and that "the system shall not be prohibited from producing negative or positive pressure." The energy code does not include requirements for system design.

Heat and energy recovery ventilation systems (HRVs and ERVs, respectively) are energy efficient and cost effective in extreme heating and cooling climates (Figure 16-5).

TABLE 16-1 Maximum run length (feet)[a]

Nominal Pipe Diameter of Largest Diameter Pipe in the Run (inch)	⅜	½	¾	> ¾
Maximum Run Length	30	20	10	5

For SI: 1 inch = 25.4 mm, 1 foot = 304.8 mm.

a. Total length of all piping from the distribution manifold or the recirculation loop to a point of use.

You Should Know

Building design offers several options for mechanical ventilation systems. Considerations for each system are summarized in Table 16-2. ●

TABLE 16-2 Comparison of Mechanical Ventilation Systems

Ventilation Type	Pros	Cons
Exhaust Only Air is exhausted from the house with a fan	• Easy to install • Simple method for spot ventilation • Inexpensive	• Negative pressure may cause backdrafting • Makeup air is from random sources • Removes heated or cooled air
Supply Only Air is supplied into the house with a fan	• Does not interfere with combustion appliances • Positive pressures inhibit pollutants from entering • Delivers to important locations	• Does not remove indoor air pollutants at their source • Brings in hot or cold air or moisture from the outside • Air circulation can feel drafty • Furnace fan runs more often unless fan has an ECM (variable-speed motor)
Balanced Air Exchange System Heat and energy recovery ventilators supply and exhaust air	• No combustion impact • No induced infiltration/exfiltration • Can be regulated to optimize performance • Provides equal supply and exhaust air • Recovers up to 80 percent of the energy in air exchanged	• More complicated design considerations • Over-ventilation unless the building is tight • Cost

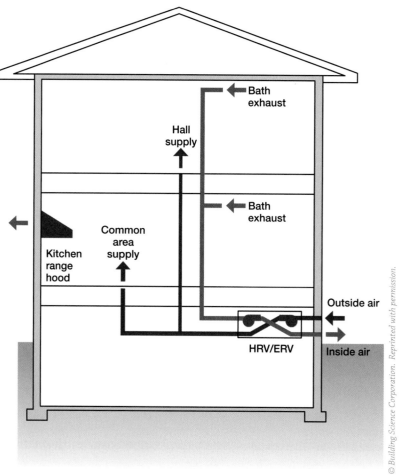

FIGURE 16-5 Simple ERV/HRV schematic layout

Heat loss is minimized by using conditioned exhaust air to warm or cool fresh incoming air. The ERV heat exchanger transfers water vapor and the heat energy. An HRV only transfers the heat energy. ERVs are most effective in warm, humid climates. HRVs (Figure 16-6) are most effective in cold, dry climates. Simple systems are easily incorporated in new building design.

Outdoor air intakes and exhausts must have automatic or gravity-operated dampers that close when the mechanical system is not in operation. The louvered vent terminations on clothes dryer and bathroom exhaust ducts are common examples of gravity-operated dampers. These comply with the provision.

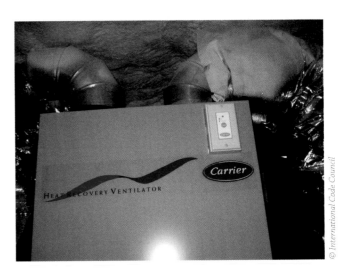

FIGURE 16-6 An HRV installed

Heating and cooling equipment sizing is important to integrating building systems thinking in new design. Properly sealed and insulated building envelopes require more consideration than "that's the way we always did it." Provisions mandated by the energy code for thermal- and air-leakage improvements allow for smaller HVAC equipment. Equipment that is correctly sized operates efficiently and saves energy costs and maintenance expense. Manual J of the Air Conditioning Contractors of America (ACCA) provides an accepted method for calculating heating and cooling loads. Check with the local code official for other approved sizing methods used in the jurisdiction.

Efficient fan motors save electric energy. *Efficacy* means "the power to produce an effect"; in this context, it refers to how many watts of electricity it takes to move a cubic foot of air per minute (CFM). Table R403.5.1 (see Table 16-3) lists fan locations and minimum efficiency requirements. Information demonstrating compliance is required to be included in notes, mechanical plans, or equipment schedules. These fans exceed the minimums for bathroom fans both less than and greater than 90 CFM.

TABLE 16-3 Mechanical ventilation system fan efficacy

Fan Location	Air Flow Rate Minimum (Cfm)	Minimum Efficacy (Cfm/Watt)	Air Flow Rate Maximum (Cfm)
Range hoods	Any	2.8 cfm/watt	Any
In-line fan	Any	2.8 cfm/watt	Any
Bathroom, utility room	10	1.4 cfm/watt	<90
Bathroom, utility room	90	2.8 cfm/watt	Any

For SI: I cfm = 28.3 L/min.

FIGURE 16-7 Snow melt at work

FIGURE 16-8 Snow melt control sensor

SNOW MELT

Snow and ice accumulation on walkways, stairs, driveways and parking areas, may create unsafe conditions around dwellings in cold climate zones (Figure 16-7). Operational controls for snow and ice systems are mandatory. Automatic controls include surface temperature sensors coupled with a moisture sensor (Figure 16-8). When the temperature is above 40°F at the surface sensors and it is not raining or snowing, the automatic control turns the system off. A manual shutoff control complies with the control provision. The manual control option requires an outdoor sensor set limiting shutoff only when the temperature is above 40°F. The energy code does not reference standards or limit the size of snow melt installations.

POOLS AND SPAS

Energy-saving provisions for residential pool and spa pumps and heaters are addressed in the IECC. Vapor-retardant covers are required to limit evaporation, but not covers specifically installed to maintain the heated water temperature. This requirement is waived if more than 70 percent of the annual energy required to heat the pool is recovered on site.

Pool heaters must have an easy-to-reach on/off switch mounted outside of the heater. Filter pumps and circulation motors are required to have control systems that automatically turn these devices on and off according to a preset schedule. Exceptions to the time switch provisions allow for public health requirements and pumps that operate pool heating systems using solar or waste heat recovery.

Electrical Power and Lighting

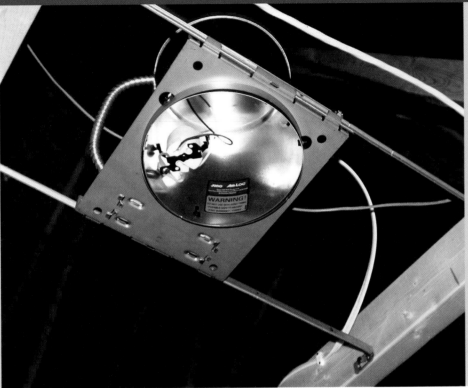

© International Code Council

Lighting uses about 10 percent of residential electrical energy. A mandatory provision requires that at least 75 percent of installed light fixtures in a dwelling must have *high-efficacy lamps*. The term *high-efficacy* regarding lighting fulfills the same energy-saving intent as the high-efficacy fan motor requirements serve in moving air. Ventilation and lighting, both necessary functions in a comfortable living environment, offer great opportunities for energy savings.

LIGHTING

The IECC defines high-efficacy lamps in lumens (light emitted) per watts used to produce the light. Lamps that produce 60 lumens/W for lamps over 40 W, 50 lumens/W for lamps over 15 W to 40 W, and 40 lumens/W for lamps 15 W or less meet the provision (Table 17-1). **[Ref. 404.1]** A lamp is just the lightbulb itself. A chandelier is one fixture but may have many lamps. Lightbulb packages include the lumen information needed to determine compliance.

TABLE 17-1 High-efficacy lamps

Lamp	Efficiency
≤ 15 W	40 lumens/W
>15 W–40 W	50 lumens/W
>40 W	60 lumens/W

© International Code Council

Example:

The information on a package of 40-watt lightbulbs indicates that the lamp produces 445 lumens. Is this a high-efficacy lamp?

Answer: 445 lumens/40 watts = 11 lumens per watt
11 lumens/watt is less than 50 lumens/watt, so this is not a high-efficacy lamp.

Most fluorescent lamps meet the high-efficacy requirement; most incandescent bulbs do not (Figure 17-1). Light from fluorescent lamps "feels" different and is often described as "cool-white" or "warm-white" light. For most residential applications, it is usually appropriate to specify warmer lamp colors (CCT = 2700–3000 K), as they give a warmer feel.

Color temperature is the standard method to describe the visible range of light, and looks like a rainbow (Figure 17-2). The color of light emitted is measured and expressed in Kelvin (K) units, and this information can be found on most lightbulb packages. The information for

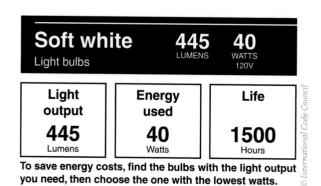

© International Code Council

FIGURE 17-1 Incandescent bulbs do not meet high-efficacy requirements

Color temperatures in the Kelvin scale

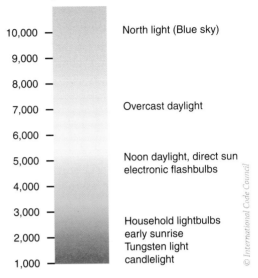

10,000 — North light (Blue sky)

9,000 —

8,000 —

7,000 — Overcast daylight

6,000 —

5,000 — Noon daylight, direct sun
electronic flashbulbs

4,000 —

3,000 —

2,000 — Household lightbulbs
early sunrise
Tungsten light

1,000 — candlelight

FIGURE 17-2 The color of light emitted is measured in Kelvin (K) units

Compact fluorescent lamp

Mini spiral	950 lumens	15 watts	10,000 hours
Incand. bulb	800 lumens	60 watts	1,000 hours

FE – IIS – 15W

15W   E26   2700K

120V 60Hz 275mA

FIGURE 17-3 Compact fluorescent lightbulb

the 15-W compact fluorescent lamp (CFL) in Figure 17-3 indicates that the light measures 2700 K and closely resembles a typical household lightbulb.

Low-voltage lighting systems generally operate at 30 volts or less. These low-voltage systems are exempt from the lighting system requirements. **[Ref. R404.1 Exception]**

You Should Know

There are two ways to calculate lighting efficiency compliance: lamp count and fixture count. The 75 percent calculation is made for all electrical lighting fixtures that are not covered by the low-voltage exception. ●

GLOSSARY

A

above-grade wall – In the energy code for residential buildings, when more than 50% of a wall is above grade it is considered an above-grade wall. The measurement is the percentage of the portion of the building foundation or other exterior wall enclosing conditioned space above or below grade. For commercial prescriptive requirements see also Section C402.2.2

accessible – Equipment located such that it is not in a locked room and is easy to approach. Readily accessible means the equipment can be maintained and inspected without crawling over other equipment or using a portable ladder. (Not related to accessible as in accessibility for the disabled.)

addition – In the energy code, an increase in the volume of conditioned space. This may be in floor area or height.

adhesive – A substance that produces a steady and firm attachment or adhesion between two surfaces.

air barrier – Material(s) that provide(s) a barrier to air leakage through the building envelope.

air changes per hour (ACH) – In a room or space, the complete replacement of air volume in a 1-hour period.

air conditioning – The mechanical process of treating air by controlling its temperature, humidity, and distribution.

air exfiltration – Uncontrolled exit of air from a room or enclosed space to outdoors.

air infiltration – Uncontrolled entry of outside air into a room or an enclosed space.

American Recovery and Reinvestment Act (ARRA) – On February 13, 2009, in response to the U.S. economic crisis, Congress passed the American Recovery and Reinvestment Act of 2009, commonly referred to as the "stimulus package."

annual fuel utilization efficiency (AFUE) – A measure of average fuel-use efficiency of combustion-type equipment such as furnaces and boilers over a year calculated as the ratio of annual output energy to annual input energy.

approved – Acceptable to the code official or other authority having jurisdiction. Approval is the result of investigation or by reason of accepted principles understood to fulfill the intent of specific code provisions.

ASHRAE – American Society of Heating, Refrigerating, and Air-conditioning Engineers.

ASTM international – Formerly known as the American Society for Testing and Materials; a globally recognized leader in the development and delivery of international voluntary consensus standards.

authority having jurisdiction (AHJ) – The official or entity with authority to enforce the jurisdiction's adopted codes and ordinances.

automatic – Self-acting, operating by its own mechanism.

B

basement wall – In the energy code for residential buildings, when more than 50% of a wall is below grade it is considered a basement wall. The measurement is the percentage of the portion of the building foundation or other exterior walls enclosing space above or below grade. For commercial prescriptive requirements see also Section C402.2.2.

Btu (British thermal unit) – The amount of heat required to raise the temperature of a pound of water 1°F. Burning a wooden kitchen match produces about 1 Btu.

building envelope – The components of the building exterior that separate the outside environment from the inside environment.

building thermal envelope – Basement walls, exterior walls, floor, roof, any other building element part of the six-side cube that provides a boundary between conditioned space and unconditioned space.

C

coefficient of performance (COP) – Of two types, cooling and heating. COP-Cooling is a measure of performance used for refrigeration systems; COP-Heating is a measure of performance used for heat pump systems.

commercial building – In the context of the energy code, all buildings that are not residential buildings.

commissioning – A process of verifying and documenting that selected building systems or equipment perform and function as originally designed and intended.

compact fluorescent light (CFL) – A lamp of a small compact shape with a single base that uses electricity to excite mercury vapor that produces ultraviolet light, producing a visible light. A fluorescent lamp converts electrical power into useful light much more efficiently than incandescent lamps; incandescent lamp light is produced by electric resistance, such as the glow of an electric oven element.

conditioned floor area – The floor area of the space or building expressed in square feet and easily found on the floor plan in building documents. Only areas within the building that are heated or cooled count as conditioned floor area.

conditioned space – An area, room, or space being heated or cooled by any equipment or appliance. In the energy code, the conditioned space may contain uninsulated ducts or an opening directly into an adjacent conditioned space.

continuous air barrier – An assembly or building material that restricts or resists airflow into or out of the building through the building thermal envelope.

continuous insulation (CI) – Insulation that is continuous across all structural members with no significant thermal breaks (thermal bridges).

cooling degree days (CDD) – The number of degrees difference each day above 65°F. If the average outside temperature is 80°F for 10 days, the cooling degree days for that time period is 150 CCD [(80°F − 65°F) = (15°F × 10 days) = 150 CDD]. The days are added together for the entire cooling season.

D

degree day (DD) – A unit of heat measurement equal to 1°F difference from a standard temperature to the average temperature of one day. The standard temperature is 65°F.

department of energy (DOE) – U.S. Department of Energy, created in 1977. The DOE is active in energy code development.

duct – A passageway usually made of sheet metal or a flexible material used to convey air throughout, into, and out of the building envelope.

dynamic glazing – Fenestration products that have the ability to change their performance properties, allowing the occupant to manually control the environment by tinting (or darkening) a window or change other features or fenestration products that change their performance automatically in response to a control or environmental signal.

E

energy – The capacity for doing work. Thermal energy is expressed in Btus; electrical energy is measured in kilowatt-hours.

energy efficiency ratio (EER) – The ratio of usable output to input of energy. The output number is Btus per hour of performance to the input number of the number of watts used by the mechanical component.

enthalpy – In the case of air, the total energy of the air including the water contained in the air. For example, conditioning "dry" air requires less energy than conditioning "humid" air.

F

fenestration – Skylights, roof windows, vertical windows, opaque doors, glazed doors, and glazed block, including products with glass and nonglass glazing materials.

flue – A pipe or shaft that contains exhaust gases and carries them to the outside environment.

H

heating degree day (HDD) – The number of degrees difference each day below 65°F. If the average outside temperature is 40°F for 10 days, the heating degree days for that time period is 250 HHD [(65°F − 40°F) = (25°F × 10 days) = 250 HDD]. The days are added together for the entire heating season.

heating, ventilation, and air conditioning (HVAC) – The mechanical system that controls the air temperature and change of air within a space or building for health and comfort.

I

IBC – *International Building Code* published by the ICC.

ICC – International Code Council, the publisher of the family of I-Codes, including the IBC, IRC, and the IECC.

IECC – *International Energy Conservation Code* published by the ICC.

Illuminating Engineering Society of North America (IESNA) – The association dedicated to improving the lighted environment; founded in 1906.

infiltration – Uncontrolled flow of air into the building, whether intentional or not, through windows, doors, cracks, and seams, or other areas that allow air to penetrate.

insulation – The material used as part of the building thermal envelope to slow the flow of heat through the enclosure. Insulation is made from a number of different materials, including glass or mineral fiber, cellulose, polystyrene, and polyurethane foam. It can be loose filled and blown into wall cavities and attic space, or installed as batts, boards, or blocks in the thermal envelope.

J

jurisdiction – The government unit that adopts codes or regulations under due legislative authority.

L

light-emitting diode (LED) – An electronic device or light source illuminated just by the movement of electrons in a semiconductor; LEDs do not get very hot.

M

manual – Capable of being operated by personal intervention.

mass wall – Masonry, concrete or similar dense building material weighing at least 35 pounds per square foot or solid wood or other less dense material weighting at least 25 pounds per square foot. These wall material types provide thermal storage benefits. Examples include adobe, filled concrete blocks, poured-in-place reinforced concrete and sawn logs.

N

National Appliance Energy Conservation Act (NAECA) – The National Appliance Energy Conservation Act of 1987 established minimum efficiency standards for many household appliances, such as air conditioners, clothes dryers, clothes washers, dishwashers, refrigerators, freezers, kitchen ranges, ovens, and pool heaters.

National Fenestration Rating Council (NFRC) – A nonprofit organization formed in response to the energy crisis of the early 1970s that administers a uniform and independent rating and labeling system for the energy performance of windows, doors, skylights, and attachment products.

O

outdoor air – Air introduced into the building from the outdoors and not previously circulated through conditioner equipment.

P

performance – A methodology based on objectives and desired outcomes, rather than the specifics of how to achieve it.

prescriptive – A methodology that provides specific steps or actions to result in a desired outcome.

projection factor (PF) – The horizontal depth of the overhang divided by the distance from the bottom of the fenestration to the lowest point of the bottom of the overhang.

R

recirculated air – Air returned from the inside of the building that passes through the conditioner before being supplied again to the conditioned space.

residential building – In the context of energy code, detached one- and two-family dwellings and multiple single-family dwellings (townhouses) as well as Group R-2, R-3, and R-4 buildings three stories or less in height (for all occupancy classifications, including R-2, R-3, and R-4, refer to the IBC).

return air – Air reintroduced into the building without passing through the conditioner.

R-value – Unit of measurement for thermal resistance of materials or assemblies. The higher the R-value of a material or assembly, the less heat or cold passes through and the more energy efficient the insulation.

S

sealant – A mixture used to fill and seal joints where some movement between materials is expected.

seasonal energy efficiency ratio (SEER) – Unit of measurement for the efficiency of air-conditioning systems calculated by dividing the total cooling output during the normal annual cooling period usage in Btu by total electric energy input during the same period in watt-hour (Wh)

solar heat gain coefficient (SHGC) – A measure of how well a window or skylight assembly blocks the heat from sunlight, expressed as a number between 0 and 1. The lower the SHGC, the more heat is reflected by the window, resulting in greater energy efficiency in hot climates.

stack effect – The upward and downward movement of air in a building due to temperature difference. Pressure caused by the air movement can lead to air leakage into and out of the building envelope.

T

thermal bridge – An element in a building or building envelope assembly that provides less thermal resistance than the adjacent construction.

tons of cooling – Air-conditioning unit cooling capacity is often expressed as "tons of AC." 12,000 Btus of cooling equals 1 ton. An 8-ton air-conditioner produces 96,000 Btus of cooling capacity.

U

UA – The thermal transmittance of a specific contact area.

U-factor – A measure of heat conduction for building elements or assemblies. It is the inverse of R-value (1/R).

V

vent – A pipe or conduit composed of factory-made components for conveying air or gasses to the atmosphere.

ventilation – The natural or mechanical process of supplying air to or removing air from a room or space.

vestibule – An intermediate space near the primary entrance doors separating the indoor conditioned air from the outside air and intended to reduce infiltration of outside air.

visible transmittance (VT) – A measure of the amount of visible light transmitted for an assembly, expressed in units between 0 and 1. The higher the VT, the more light is transmitted and the more efficient is the use of daylight.

W

watt – The unit of measurement for electrical power.

weather-stripping – The process of sealing or the material used to seal around openings such as doors and windows to retard the infiltration or exfiltration of air.

INDEX

Information in figures and tables is denoted by *f* and *t*.

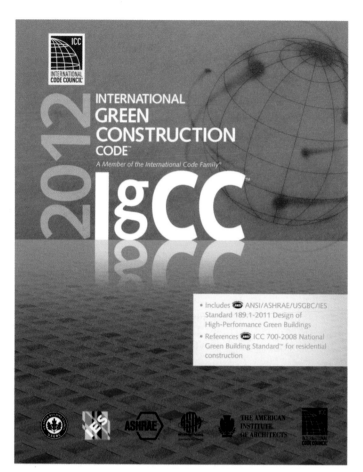